The Callendar Effect

Guy Stewart Callendar in 1934, about the time he turned his attention to the CO_2-climate question.

The Callendar Effect

The Life and Work of Guy Stewart Callendar (1898–1964), the Scientist Who Established the Carbon Dioxide Theory of Climate Change

JAMES RODGER FLEMING

American Meteorological Society

The Callendar Effect: The Life and Work of Guy Stewart Callendar (1898–1964), the Scientist Who Established the Carbon Dioxide Theory of Climate Change

Published by the American Meteorological Society
45 Beacon Street, Boston, Massachusetts 02108

Also available from AMS Books: *The Papers of Guy Stewart Callendar*, Digital Edition on DVD, James Rodger Fleming and Jason Thomas Fleming, Eds. (Boston: American Meteorological Society, 2007). This research-quality digital archive includes Guy Stewart Callendar's manuscript letters, papers, journals, documents, and family photographs.

For a catalog of AMS Books, see www.ametsoc.org/pubs/books. To order, call (617) 227-2426, extension 686, or email amsorder@ametsoc.org.

Library of Congress Cataloging-in-Publication Data

Fleming, James Rodger.
   The Callendar effect : the life and times of Guy Stewart Callendar, the scientist who established the carbon dioxide theory of climate change / James Rodger Fleming.
      p. cm.
   Includes bibliographical references and index.
   ISBN-13: 978-1-878220-76-9 (alk. paper)
   ISBN-10: 1-878220-76-4 (alk. paper)
      1. Callendar, Guy Stewart, 1898-1964. 2. Scientists—Great Britain—Biography.
3. Engineers—Great Britain—Biography. 4. Atmospheric carbon dioxide. 5. Global warming. I. Title.

   Q143.C255F54 2007
   509.2—dc22
   [B]

                                                                    2006029902

Printed in the United States of America by Kase Printing, Inc.

Dedicated to Anne and Bridget,
twin daughters of Guy Stewart Callendar

Contents

Illustrations

Acknowledgments

Research support for this book and the digital archive was provided by the National Science Foundation under Grant SES-0114998. Any opinions, findings, and conclusions or recommendations expressed in this material are those of the author and do not necessarily reflect the views of the National Science Foundation.

G. S. Callendar's daughter Bridget provided precious family photos, stories, letters, and encouragement that made the story come alive.

Callendar's scientific letters and notebooks were supplied by Professor Phil Jones, Director of the Climatic Research Unit, and Alan Ovenden, Map Librarian at the University of East Anglia, Norwich.

NSF and my home institution, Colby College, funded several extended research trips to England and supported the work of scanning the documents and organizing and preserving the archival collection. These immense tasks were spearheaded by my sons Jason Thomas Fleming (organization and scanning) and Jamitto David Fleming (research and preservation). Thanks are also due my adminstrative assistant, Alice Ridky, and several student assistants, Anthony Abakisi, Madeline Horwitz, Jessica A. Foster, and Derek Snyder. The last two in particular took on major tasks involving research and interpretive writing.

Early versions of this work were presented in seminars and presentations at the University of East Anglia, The University of Arizona, Scripps Institution of Oceanography, and at annual meetings of the American Meteorological Society, the American Geophysical Union, Atmospheric Science Librarians International, and the Society for the History of Technology. I thank Martin Collins and Marc Rothenberg of the Smithsonian Institution for their insightful and helpful comments on a penultimate version of the text.

In alphabetical order of instutution, I would like to thank the following for their assistance:

British Library Newspaper Library, London; British Meteorological Office Library, Steve Jebson, Librarian; British National Archives and Public Record Office; Durston House School, Peter Craze, Head Master and Henry Ross, IT Manager; Ealing Lawn Tennis Club, Eric Leach, Archivist; Ealing Library, Dr. Jonathan Oates; George Mason University, Professor Josh Greenberg; Horsham Council of Churches, Rev. Philip Dale; Imperial College of Science, Technology and Medicine, London, Anne Barrett, College Archivist, and Hilary McEwan, Special Collections Archivist; Imperial War Museum, Katharine Martin; Institution of Mechanical Engineers, Keith Moore, Archivist, and Fenella Philpot, Archives Officer; Royal Meteorological Society; Royal Society of London, Joanna Corden, Archivist, and Clara Anderson, Assistant Archivist; London Science Museum, Jane Insley and Neil Brown, Curators; Scripps Institution of Oceanography, Deborah Day, Archivist, and Rich Kacmar, Assistant Archivist; St. Paul's School, Simon May, Archivist; University of Dundee, Matthew Jarron, Curator of Museum Services; University of East Anglia, Professor Mike Hulme, Executive Director, Tyndall Centre for Climate Change Research, Professor Peter Brimblecomb and Dr. Francis Mudge, School of Environmental Sciences.

Production of the book and DVD were professionally facilitated by the staff of the American Meteorological Society: Executive Director Keith Seitter, Historical Monographs Editor John Perry, Director of Publications Ken Heideman, and Books and Monographs Manager Sarah Jane Shangraw.

Introduction

Guy Stewart Callendar (1898–1964) is noted for identifying, in 1938, the link between the artificial production of carbon dioxide and global warming. Today this is called the "Callendar Effect." He was one of Britain's leading steam and combustion engineers, a specialist in infrared physics, author of the standard reference book on the properties of steam at high temperatures and pressures, and designer of the burners of the notable World War II airfield fog dispersal system, FIDO. He was keenly interested in weather and climate, taking measurement so accurate that they were used to correct the official temperature records of central England and collecting a series of worldwide weather data that showed an unprecedented warming trend in the first four decades of the twentieth century. He formulated a coherent theory of infrared absorption and emission by trace gases, established the nineteenth-century background concentration of carbon dioxide, and argued that its atmospheric concentration was rising due to human activities, which was causing the climate to warm.

Callendar's contributions to climatology led the way in the mid-twentieth-century transition from the traditional practice of gathering descriptive climate statistics to the new and exciting field of climate dynamics. In the first half of the twentieth century, the carbon dioxide theory of climate change had fallen out of favor with climatologists. In the 1930s Callendar revived and reformulated this theory by arguing that rising global temperatures and increased coal burning were closely linked. Employed in defense research, and working from his home in West Sussex, Callendar compiled weather data from frontier stations around the world that clearly indicated a global warming trend in the early decades of the twentieth century. He also documented the retreat of glaciers. To support his hypothesis, he calculated the world use of fossil fuels, estimated carbon reservoirs and uptake by the biosphere and oceans, and compiled historical measurements that showed rising concentrations of carbon dioxide in the atmosphere. Callendar established the now standard number of 290 parts per million (ppm) as the background concentration of carbon dioxide in 1900 and estimated there had been a ten percent increase by 1938. Based on new work on the infrared spectrum and calculations of the absorption and emission of radiation by trace gases in the atmosphere, Callendar established the CO_2 theory of climate change in its recognizably modern form, reviving it from its earlier, physically unrealistic and moribund status. He concluded that the rising temperature trend was due to anthropogenic increases in the concentration of atmospheric carbon dioxide, primarily through the processes of combustion.

Until recently, G. S. Callendar had been largely overlooked by historians and scientists. He was a quiet, family oriented man, an avid sportsman, supremely competent, widely published and cited, yet unassuming. He received few special honors, held no academic appointments, and left relatively few letters and no personal journals. His scientific papers, although considered valuable, were in danger of being scattered, damaged, and lost. No one in the climate research community had any photographs of the man or his family. Under these circumstances, preparing his biography and a digital archive of his papers were daunting tasks. Callendar's life was reconstructed from a variety of sources including scientific correspondence and notebooks, family papers (graciously provided by his daughter Bridget), and the well-documented life of his father, the famous physicist Hugh Longbourne Callendar, who nurtured his son's career and introduced him to the technical elite of England.

Callendar was a Fellow of the Royal Meteorological Society and served on its council. He was also a Fellow of the Glaciological Society. He counted many distinguished scientists as his friends, colleagues, and coworkers. He received a first-rate technical education and entered into collaborations with

Britain's technical elite on steam research and the infrared spectra of complex molecules. His work on the thermodynamics of steam was foundational for steam-plant design calculations in Great Britain for more than three decades. His papers on the infrared properties of trace gases drew rave reviews from leading meteorologists and climatologists and influenced the later development of the field. His work in defense-related research in two world wars and the cold war was directed, wherever possible, toward non-violent ends.

The legacy of Guy Stewart Callendar continues through the Callendar Effect, influencing scientists of his own and subsequent generations and now historians of science to examine the paths by which we have arrived at our current state of climate knowledge and apprehension.[1] This volume provides an overview of his life and times and introduces an optional digital archive of the Callendar Papers. I hope that it provides the reader with deeper perspectives on an earlier era in science and engineering, allows a glance into Callendar's personal life and accomplishments, and stimulates both broader awareness and further research on the life and work of this unpretentious, but creative and fascinating individual.

The Callendar Effect: Climatic change brought about by [anthropogenic] increases in the concentration of atmospheric carbon dioxide (CO_2), primarily through the processes of combustion. The actuality of such changes was proposed in 1938 by the English scientist G. S. Callendar, son of H. L. Callendar. *See also* greenhouse effect.

—*Encyclopedia Britannica*

CHAPTER ONE

The Early Years to 1930

The Callendar boys shared their father's love for science and engineering, while sister Cecil cultivated the arts.

—Leslie Callendar

Guy Stewart Callendar was strongly influenced by his famous father, who nurtured his son's scientific and extracurricular interests and served as his mentor. Hugh Longbourne Callendar, first son of the Reverend Hugh Callendar and Anne Cecilia Longbourne, was born April 18, 1863, in Hatherop, Gloucestershire and was christened May 24, 1863, in his father's church.[1] Following his father's untimely death in 1867, young Hugh was nurtured by both the Callendar and Longbourne families and developed a very protective attitude toward his mother and siblings.

From early childhood Hugh was a self-motivated learner: precocious in languages and mathematics, with a knack for technical matters. His skills were cultivated at a young age by a private tutor.[2] By age 10, he had constructed a number of technical devices, including a Rhumbkorff induction coil, a Wimshurst electrostatic generator, and a telegraph network that he used for communication around the house (having taught himself Morse code).[3] At age 11, with the assistance of his maternal uncle J. V. Longbourne, Hugh entered Marlborough College. Hugh's interests and hobbies included astronomy, nature study, competitive shooting, gymnastics, football, and handicraft. Some of the younger boys at school believed that he controlled the weather because he was put in charge of the barometer in the house.[4] He constructed an automatic wind recorder and designed a fountain pen using an angular glass tube and a rubber stopper. His mother liked the pen so much, she used one for her letter writing for more than 60 years. He also devised a system for testing sight and color blindness.

Cambridge Years

Hugh Callendar entered Trinity College, Cambridge in 1882. During his university years, Callendar reportedly studied 10 to 12 hours a day yet still reserved two hours for afternoon sporting—he excelled in tennis, gymnastics, lacrosse, and shooting. To boost his efficiency, he devised a method of speed writing, subsequently known as the Cambridge system of shorthand, teaching the system to his friends and associates, as well as publishing several books about it.[5] Earning first class honors in Classics (1884) and Mathematics (1885), Callendar began experimental work in physics at the Cavendish Laboratory with Professor J. J. Thomson.[6] The two men greatly influenced each other. Thomson learned Callendar's shorthand system and used it for note taking throughout his life. Callendar followed Thomson's suggestion to study the problem of metallic resistance thermometry, a decision that shaped the rest of his career.

The Platinum Resistance Thermometer and Related Inventions

In 1885, after eight months of extensive experimentation and research in cramped quarters at the Cavendish Laboratory (he used a windowsill as a workbench), Callendar developed an accurate platinum resistance thermometer suitable for high temperature measurements (Figure 1.1). Compared to earlier devices, it had an extended temperature range and held its calibration over time. According to Thomson, Callendar "gave to physics a new tool to determine temperature with an ease and accuracy never before obtainable. . . ."[7]

In Callendar's original design, coils of fine platinum wire were wound on thin mica plates connected to silver leads. He added annealed pure platinum contacts and corrected for errors made by other scientists in the relationship between temperature and resistance. Callendar presented his findings to the Royal Society of London on June 10, 1886, and patented his thermometer the following year.[8] Arguably the most significant invention of his career, Callendar's new thermometer was able to measure temperature

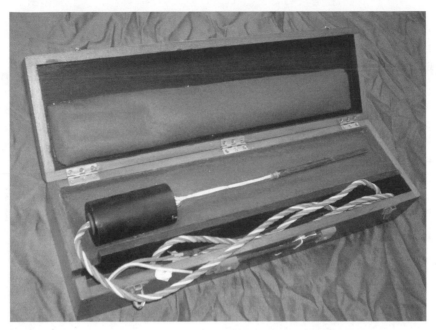

Fig. 1.1. Platinum Resistance Thermometer designed by H. L. Callendar. Image courtesy of the University of Dundee.

with great precision from −190° to 660°C; later models were useable up to 1200°C. Callendar employed this instrument in many of his experimental investigations. The new platinum thermometers most accurately measured the melting point of a variety of metals and allowed for further work in new metal alloy construction. The Cambridge Instrument Company produced a commercial version of the Callendar platinum thermometers, which found immediate applications in metallurgical engineering.[9]

In his investigations of the properties of steam Callendar used his sensitive platinum resistance thermometer to follow rapid variations of temperature without appreciable lag. As explained in the London Science Museum *Catalogue*:

> The instrument, in use, is mounted on the piston of a steam engine, and readings are taken at any desired point of the stroke by means of a periodic contact mounted on the revolving shaft. With this arrangement Callendar was able to examine carefully the relation between pressure and temperature during the adiabatic compression and expansion of steam.[10]

Callendar invented and improved myriad scientific devices related to precision thermometry. Many of these instruments were presented to or loaned to the London Science Museum by Imperial College of Science and Technology. For example, Callendar and E. H. Griffiths developed a modified form of Wheatstone bridge to measure the resistance of a heated platinum wire and thus indicate its temperature.[11] With this improved device, used in conjunction with the platinum resistance thermometer, it was possible to measure small changes in resistance with accuracy correct to one-tenth of a degree.[12] Callendar devised a platinum resistance thermometer only 0.025-milimeter thick for biological research applications. The Cambridge physiologist W. H. Gaskell used this device in 1887 to detect heat generated by the beating of a frog's heart.[13] For clinical use, Callendar invented medical thermometers connected to automatic temperature recorders, which could be used rectally or held in place under a patient's armpit.[14] Callendar also designed an automatic data recorder to convert the electrical signal from any instrument into the movement of a pen on a clockwork-driven cylinder of paper.[15]

In 1886 Callendar was elected a Fellow of Trinity College, Cambridge. Acceding to the wishes of his family, he tried his hand briefly at law and medicine before accepting a position in 1888 as professor of physics at Royal Holloway College, a recently opened college for women in Egham Hill, Surrey. He held this position for the next five years.[16] In addition to his teaching duties at Royal Holloway, Callendar vigorously continued his research at the Cavendish Laboratory. In 1894 he extended the heat range of his platinum resistance thermometer to 1600°C by adding resistances into the circuit and using a galvanometer as a sensitive temperature indicator.[17]

Sojourn in Canada

In 1893 Callendar was appointed professor of physics at McGill University in Montreal, Canada, with his office and laboratory in the newly constructed and state-of-the-art Macdonald Physics Building. Before moving to Canada he asked for the hand of Victoria Mary Stewart, the eldest daughter of Alan Stewart of Saundersfoot, Pembrokeshire. He had met the attractive, energetic redhead at Cambridge during his fellowship years. Callendar returned to England in the spring of 1894, married Victoria, and brought her back to Canada with him.[18] That same year he was elected a Fellow of the Royal Society of London.

Callendar was a formidable presence in the McGill physics department. He worked on the thermodynamics of the steam engine and developed a method for determining the total heat of steam using a new type of calorimeter. He designed the first X-ray experiment in Canada and on February 7, 1896, produced satisfactory photographs.[19] His device was used at a local hospital to generate medical X-rays of patients with embedded needles, swallowed coins, lung cavities from pneumonia, bullet wounds, kidney stones, broken bones, and fractured skulls. With C. H. McLeod, director of the McGill Observatory, he conducted meteorological investigations, studying the variation of soil temperature with depth. Callendar lectured frequently at McGill and attracted scientists from around the world to hear him speak.[20]

Honors continued to accrue to Callendar. For his work on theoretical and practical aspects of steam engineering, he received the James Watt Medal and the Telford Premium from the Institute of Civil Engineers. He was elected a Fellow of the Royal Society of Canada in 1896 and was awarded an honorary Doctor of Laws by McGill University in 1898.[21] Outside of the laboratory and classroom, Professor Callendar maintained his athletic interests, representing McGill in tennis, lacrosse, and shooting.

Enter Guy Stewart Callendar

Guy Stewart Callendar was born on February 9, 1898, in Montreal, Canada. His older siblings Cecil and Leslie Hugh were also born in Montreal. Before Guy's first birthday the family returned to England, his father having been offered the prestigious Quain Chair of Physics at University College, London. Although McGill made an enticing counteroffer, nearly doubling his salary and adding an additional professorship in astronomy, Hugh decided his prospects were brighter in England. Ernest Rutherford, the new chair of physics at McGill, had this to say about his predecessor: "Callendar here was considered a universal genius and I gain a sort of reflected glory by carrying on with things Callendar alone was able to do. The trouble is that Callendar left such a reputation behind him that I have to keep rather in the background at present."[22] The Callendar family too had made a distinct impression on the community. On the eve of their departure approximately 200 people came to bid them farewell.[23]

Cecil, Guy, and Leslie are depicted in Figure 1.2, with their mother Victoria.

The children appear to be wearing souvenir deck hats from their recent Atlantic crossing.[24] Another early image of Guy with his maternal grandfather, Alan Stewart of Saundersfoot, Pembrokeshire (Figure 1.3) was taken in 1901, perhaps following Guy's baptism at St. Issells Church in Saundersfoot.[25]

In 1902, H. L. Callendar moved from University College across town to South Kensington to accept a professorship of physics at the Royal College of Science which became part of Imperial College in 1907. Here he worked for the rest of his life.

Motoring Interests

Guy Callendar and his siblings benefited from the motoring interests of their father. H. L. Callendar spent his free time tinkering with cars, racing them, or taking his family for high-speed jaunts around the English countryside. Perhaps the stereotype of the tinkering physicist as motoring enthusiast began here. Around 1898 Callendar conducted research on the internal combustion engine, worked out a formula for motor horsepower, and immediately became intrigued with motorcycling.[26] In 1902 he purchased one of the first Clement-Gerrard motorbikes in England. He added an eight-speed gearbox, a cooling system, and fitted an armchair and extra wheel on the front, turning the two-wheeled contraption into a supercharged, two passenger tri-cycle, allowing Callendar and his wife to tour together and climb with ease the "fearful" hills of Porlock in the West Country. Guy and his siblings were treated to shorter joy rides around the neighborhood. Leslie recalled, "I well remember being driven at 13 mph in this armchair, exceeding the speed limit then in force."[27] The tri-car was subject to frequent breakdowns and it would be up to Callendar and local blacksmiths to perform repairs.

In 1904, after he and his wife were knocked over in their tri-car by a horse, Callendar decided he needed a larger and sturdier vehicle. He purchased an eight horsepower De Dion Bouton car and modified it to accommodate five passengers.[28] Callendar's next large motoring investment was a Stanley Steam Car, purchased in 1907. Being an inquisitive scientist and steam engineer, Callendar was keenly interested in its power source. Despite his initial fascination, the vehicle's steam engine proved to be dangerous and cumbersome. During a particularly catastrophic family trip to Hindhead, the steam pressure fell off, the car came to a standstill, and one of the boiler tubes burst,

Fig. 1.2. "Peeling a pear." Prize photo taken by Hugh Callendar in the summer of 1900 at "Aunt Ethel's garden in East Grinstead." Guy, age two, is sitting at his mother's feet, brother Leslie, four, is standing, and sister Cecil, five, is sitting in the tub.

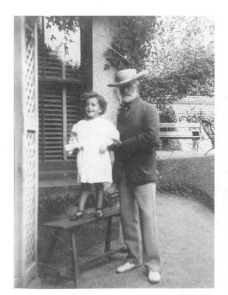

Fig. 1.3. Guy with his maternal grandfather, Alan Stewart, ca. 1901.

Fig. 1.4. Hugh Longbourne Callendar, ca. 1910.

pouring hot steam onto the passengers. Consequently, Callendar decided to trade the steam drive for a hot-air system that used a petroleum engine as a compressor. Although this adjustment slowed the overall speed of the vehicle, his custom air-drive Stanley Steam Car proved to be safer and no longer required adding water every 20 minutes.[29] These adjustments made the vehicle suitable for family touring throughout England.[30]

Ealing

Callendar's university professorship and royalties from his successful patents provided a comfortable foundation for the family, allowing them to move, in 1905, to the fashionable town of Ealing. Their four-story, 22-room home at 49 Grange Road (Figure 1.5) was covered by flowering wisteria on the front and Virginia creeper on the side and back. Facilities included two garages (or "motor houses") equipped with a pit, crane, lathe, and all the tools needed to maintain and modify the vehicles; a greenhouse; a full-size tennis lawn (also used for cricket and football); a putting green; mature trees, arbors, and a flower garden. The Callendars kept a staff of four to six servants, a chauffer, and a gardener.[31] This estate was the primary household where all four Callendar children (the youngest son Max was born in 1905) grew up. The town of Ealing had large open green spaces and numerous tennis courts. Walpole Park, with its 30 acres of trees and grass, was only a few hundred yards from Grange Road.[32] Ealing also was quite conveniently connected to central London by a new electric rail line.

According to brother Leslie, the Callendar boys shared their father's love for science and engineering, while sister Cecil had a flair for writing and painting and cultivated the arts.[33] The large library, natural history specimens, and collection of scientific instruments stimulated their curiosity. Callendar had converted the greenhouse into a laboratory where the children were encouraged to play and tinker—but not without risk. Eldest son Leslie destroyed it one day while attempting to create TNT.[34] Earlier, Leslie had accidentally blinded his brother Guy at age five, by sticking a pin in his left eye.[35]

Domestic life was regimented in the Callendar household under the supervision of both parents. Although Hugh was the breadwinner of the family, Victoria justifiably provided the backbone, managing the family while he was at the university. The boiler had to be stoked at 6:30 A.M., the dinner

Fig. 1.5. The Callendar family home from 1905 to 1930 at 49 Grange Road, Ealing. Photograph of a large pen drawing done by Cecil Callendar, age 19, in 1914.

table perfectly set, and staff to bed by 10:00 P.M. As a father, Callendar was respected by his children. Leslie writes, "His word was law among us."[36] Leisure activities included tennis and (of course) motoring. Callendar also built a rapid-shutter camera that probably took many of the family's early photos.[37]

Guy's Education

From 1909 to 1913, Guy Stewart Callendar attended Durston House School in Ealing (Figure 1.6), a small, elite, and very strict private day school founded by the brothers B. C. and R. M. Pearce. For several years the Callendar children must have walked together to school, a distance of about 1 kilometer from their home. Leslie graduated from Durston House in 1911, Guy in 1913, and Max in 1921.[38]

In the spring of 1913, Guy matriculated as a day student at the exclusive St. Paul's School in Hammersmith, a school steeped in tradition—it was founded in 1509 by John Colet, dean of St. Paul's Cathedral, and counted John Milton, Edmund Halley, and Samuel Pepys among its venerable graduates. The curriculum at St. Paul's emphasized the study of Greek, Latin, and the humanities, but a new science block had recently been opened, and athletic competition was considered compulsory.

The outbreak of the Great War had an immediate impact on Guy, in part because of the militarism in evidence at St. Paul's. Instructors left for military service, instructional hours were shortened to avoid air raids, rationing affected the dining halls, and a battlefield map appeared in the library. Military drills, including bayoneting practice, were held on campus, at Runnymede, and in Richmond Park.[39] Such public displays of patriotism did not appeal to Guy. He was also reeling from the emotional shock of losing his beloved sister Cecil in 1914 at age 19 after she contracted pneumonia. Many boys left school earlier than normal at this time, and among them was Guy Callendar, in 1915.[40] Guy's aversion to violence was learned from his father. As told by his brother Leslie, a student at Imperial College in May 1915, "the morning after the *Lusitania* had been sunk by a submarine with the loss of many men, women, and children . . . my father opened his lecture by expressing his horror with the deepest emotion I have ever heard in his voice. He then looked slowly round the students as if these young men too would soon be lost in the war, and with difficulty went on with his lecture."[41]

The loss of his left eye in a childhood accident rendered Guy unsuitable for front line service, yet he made valuable technical contributions to the war effort. For two years he worked in his father's laboratory at Imperial College as an assistant to the X-ray Committee of the Air Ministry testing a variety of apparatus, including aircraft engines at the Royal Aircraft Factory (later Establishment) in Farnborough.[42] X-ray technology allowed him to look for hairline cracks and other defects in the engine blocks. This work also

Fig. 1.6. Durston House School, 12 Castlebar Road, Ealing, ca. 1911. Image courtesy of Durston House School.

provided him with a thorough introduction to the electromagnetic spectrum. Later Guy enlisted in the Royal Naval Volunteer Reserves,[43] attaining the rank of sublieutenant (Figure 1.7). He served as a hydrophones officer and gained experience with electrical apparatus developed by his father for sound ranging and detection of submarines.[44] Since his father served as a consultant to the Board of Inventions during the war, Guy would have been exposed to a very wide range of technical ideas—not all of which were feasible—and the basis for evaluating them.

Steam

In 1919 Guy Callendar entered City and Guilds College, part of Imperial College in South Kensington, London's "scientific and cultural heartland."[45] His father was a distinguished professor at Imperial and chair of the physics department from 1908 to 1929.[46] Guy earned a certificate in Mechanics and Mathematics in 1922 and immediately went to work as a researcher in the physics department, assisting his father with problems of steam engineering.

The first edition of the famous *Callendar Steam Tables* had appeared in 1915 and subsequent improvements in steam technology required that they be continually updated. External funding from the Air Ministry and the British Electrotechnical and Allied Manufacturers' Association (BEAMA) and a special appointment from the British Electrical and Allied Industries Research Association (BEAIRA, also known as ERA) allowed Professor H. L. Callendar to hire two research assistants; one of them was Guy Stewart Callendar.[47] Guy undoubtedly helped his father prepare the 1922 and 1927 editions of the *Callendar Steam Tables*. In 1926, Guy authored his first scientific article with his father on the total heat of steam. Their research, using a new apparatus, extended the measurements to more than double the previous values—to pressures and temperatures at the critical point for water and steam. Based on this work, professor Callendar proposed three simple thermodynamic equations, of relevance to both the scientist and to the engineer, in an attempt to define the properties of steam for international standard purposes.[48]

Guy worked for eight years as an apprentice to his father, Britain's leading steam engineer. He mingled with the world's technical elite when he attended the first International Steam Table Conference in London in 1929 with his father who was a delegate. On January 21, 1930, Hugh Longbourne Callendar, age 66, died at his home in Ealing of pneumonia.[49] As a testament to Guy's significant contribution and involvement in steam research, his father had bequeathed to him "the copy right of and all income royalties and other

Fig. 1.7. G. S. Callendar, 1918, Sub Lieutenant, Royal Naval Volunteer Reserves.

Fig. 1.8. H. L. Callendar, age 60, at Dinas Powis, South Wales, 1923.

benefits which may accrue" from the Steam Tables and Diagrams, as well as all manuscripts, papers, and illustrations that are connected to them.[50] After his father's death, Guy was prepared to don the mantle of Britain's premier steam engineer, expanding and correcting the *Callendar Steam Tables* they had developed together. Guy continued his steam research until 1942 under the patronage of BEAIRA in collaboration with Alfred Egerton (see Chapter 3).

Father and Son

Guy Stewart Callendar was raised in a loving and supportive family, surrounded by curious and competitive siblings, in a household filled with books and a vast array of technical gadgets. He received a first-rate education at St. Paul's School and City and Guild's College. His father was deeply involved in scientific circles, serving as President of the Physical Society from 1910 to 1912, then as President of Section A of the British Association for the Advancement of Science, and throughout his career as an active member of the Royal Society of London. For a number of reasons, Guy Stewart Callendar's life and career followed strikingly similar paths to those established by his father.

Both father and son excelled in sports. Guy was a talented tennis player and avid cyclist. He was "runner-up" in the gentlemen's single's championship for 1928 at the Ealing Lawn Tennis Club. His older brother had equaled this accomplishment five years earlier, while his younger brother, Max, won the championship outright in 1926, 1929, and 1931.[51] Of course, Guy had competed with only one good eye! In addition, father and son shared a strong interest in motor sports. Guy received his first motorbike from his father at the age of 14 (Figure 1.9). Guy's technical engagement with internal combustion engines was probably stimulated by these early experiences and his access to a well-equipped garage.

The two men shared a profound sense of duty to country and a common cause in the Great War, accompanied by a deep aversion to violence. Both

Fig. 1.9. Guy on his motorbike, ca. 1920.

were well connected with the elites of science and engineering. Both served the British war effort in research related to technical issues, quality control, and the detection of enemy submarines.

Guy's interest in meteorology also received a boost from his father's inventions related to the atmosphere: a sunshine receiver that accurately measured and recorded total solar radiation, an electrical air temperature grid to measure and record variations in atmospheric temperature, an absolute recording bolometer, a thermopile to measure the radiation from the sun's corona, and a disc radio-balance to measure the amount of radiation emitted by the sun or other strong sources.[52] Other themes in the later career of G. S. Callendar, including research on steam, combustion, and infrared radiation, find antecedents in the work of his physicist father.

Finally, it seems that father and son were similar in personality. Both men were intensely private; both emphasized their research over their social lives; both could be accurately described as soft-spoken, detail-oriented, and hard working. In addition to being brilliant and insightful scientists, both were passionate and dedicated family men.

CHAPTER TWO

A Family Man

To "my dearest Phyllis . . . from your ever loving Guy."

Guy Stewart Callendar was a loving husband and devoted father of two. By all accounts, he enjoyed a fulfilling family life, providing his dependents with a peaceful and secure home during the worldwide depression of the 1930s, the ordeal of World War II, and Britain's post-1945 economic decline. He was an accomplished tennis player, avid bicyclist, and creative gardener. He was home often, as attested by his remarkable unbroken series of weather observations, beginning in November 1942 and extending through September 1964.[1] This was possible because his work for the government at Langhurst was near his home, within bicycling distance in good weather. After his trips to Germany in 1930 and America in 1934 to attend the International Steam Table Conferences (see Chapter 3), he took no more international trips. On evenings and weekends, and just about full time after his retirement in 1958, he pursued his weather and climate studies—his beloved "figs."—that is, in addition to being a good family man and avid sportsman.

Marriage

On August 30, 1930, Guy Stewart Callendar, age 32, married Phyllis Burdon Pentreath, age 31, at the Parish Church of Upper St. Leonards-on-Sea. The

Fig. 2.1. Guy's wedding portrait taken in 1930 in St. Leonards-on-Sea.

Fig. 2.2. Phyllis and Guy on their wedding day in 1930.

marriage certificate indicates that the groom was a "bachelor, scientist" then living in Ealing at 49 Grange Road, and the bride an unemployed "spinster" from St. Leonards-on-Sea. Her father, Harry Pentreath, was a retired civil servant. The Rev. E. Y. G. Hunter was the presiding minister. Official witnesses included Guy's brother Leslie, Phyllis's sister Anne, and I. H. Cole.[2] Although Guy is probably standing one step above his new bride in the photograph taken on the church steps, he was still an imposing 6'2" tall and she stood about 5'4" (Figure 2.2).

Early evidence from the "love letters" (Appendix B) indicates the marriage got off to a good start. This is the only documentary evidence remaining, however. Guy did not keep a personal journal or travel far from home after 1934, so there was no need to write, and Phyllis left no letters or diaries. Guy and Phyllis shared common interests in tennis and gardening. Their daughters recall a happy home life punctuated by frequent moves and periods of financial insecurity, especially in the first decade, before the family settled in Horsham. The couple celebrated 34 anniversaries together.

The St. Leonards and Hastings area, about 70 miles southeast of London, was widely known as a first-class seaside resort and "one of the most delightful watering places in the Kingdom."[3] It boasted a mild and healthful

Fig. 2.3. Guy playing tennis, 1931.

climate and romantic seaside vistas and exposures. An unbroken esplanade extended along the coast for a distance of three miles, serving to connect the pier, marina, swimming baths, theatre, concert hall, the Royal Victoria Hotel, and the East Sussex Club.

In the week leading up to his marriage, Guy would have noted with some interest the major heat wave southern England was experiencing. According to the *Middlesex County Times*, on Tuesday the thermometer reached 90°F; the following day was recorded as the hottest August 27th for 90 years, while Thursday, with a high of 95°F, was the highest temperature recorded in the last ten days of August since 1906. Friday's temperature again reached 95°F before the heat wave broke late that evening with violent thunderstorms over London and Ealing.[4] On the wedding day, the temperature in Ealing was a seasonal 82°F, cooler at the coast, with everyone agreeing that Sunday, with a high of 78°F, was "delightful." Following the wedding the couple took up residence in Ealing.

Guy and Phyllis are suspected to have first met at the Ealing Tennis Club.[5] Guy loved to compete in tournaments and often gauged the weather by its suitability for tennis (Figure 2.3).[6] Although Phyllis was not as serious as Guy about the sport, she frequently played as a young girl, and even in her later years would occasionally pick up a racket. She especially enjoyed playing badminton and participated in organized matches.[7]

A Growing Family
On November 19, 1931, the couple was blessed with identical twin daughters. Anne arrived at 3:45 P.M. and Bridget just 30 minutes later.[8] At that time the family lived at 6 North Avenue, Ealing. Guy's brother Leslie was chosen as Anne's godfather and Guy's friend and mentor Alfred Egerton (see Chapter 3) was chosen as Bridget's godfather.

When the twins were toddlers, the family moved from Ealing to Worthing, a town on the southern coast of England about 60 miles from London. Their first home was on Gerald Road, just 200 meters from the sea. Soon, they moved to Ardale Close, a cul-de-sac just around the corner. Phyllis enjoyed knitting and sewing, and would make hats and jumpers for the twins. In early family photographs, they appear to be dressed identically (Figure 2.4). The Callendar family enjoyed living near the sea and took frequent walks along the beach (Figure 2.5).

When the twins were just two months shy of their third birthday, Guy traveled to America to attend the International Steam Table Conference (see Chapter 3). Guy loved the girls dearly, playfully referring to them as "two troublesome twins," and proudly displaying their photograph to friends and colleagues aboard the ship and at the conference. While Callendar profited from his time away, he longed to return home to his family. He wrote letters

Fig. 2.4. Callendar family, 1933. Photo probably taken in Worthing. The twins are about one year old. Note that the height difference between Guy and Phyllis is reversed.

Fig. 2.5. Guy and the twins wading in the sea near their home in Worthing in 1934.

Fig. 2.6. Guy with twins on the beach on Isle of Wight, 1937, possibly near Knock Cliff. The girls are displaying treasures they collected. Note the fossil seekers along the cliffs.

to Phyllis each day addressing her as "my dearest Phyllis" and closing, "from your ever loving Guy." He suffered incredible homesickness while away and longed throughout the entire trip to return to her and the children. He wrote, "I am missing you and the babies quite a lot, it is fortunate there is so much to do." Phyllis sent him a telegram that reached him in Cork, Ireland, and a letter that arrived in New York. Guy was delighted to discover the "nice little note you left in my bag." Unfortunately, there are no letters from Phyllis in the collection. The love letters from Guy to Phyllis, written in September 1934 (Appendix B) are the only truly intimate record of Guy's emotions.[9]

In 1937 and 1938 the Callendar family vacationed in the town of Shanklin on the Isle of Wight. The island, a short ferry ride from the British mainland, just 50 miles from Worthing, offered exquisite beaches, esplanades, scenic cliffs, fossils, and interesting flora and fauna (Figure 2.6).

The War Comes to England

On September 15, 1940, the Callendars moved to their third home in Worthing, on Hurst Avenue, where they could provide a room for Phyllis's father, Harry Pentreath. Phyllis's mother had recently passed away and Harry, who was approximately 90 years old at the time, welcomed their support and company. The move to their new home proved to be especially dramatic when, on moving day, the family witnessed a major engagement of the Battle of Britain in the skies directly above them. Since early July, Germany had been trying to gain control of the straits of Dover. Press and radio reports were filled with news about German air raids and Royal Air Force (RAF) resistance. Cities in southern England were being hit and night raids were increasing in frequency and intensity. London was targeted on September 7.

At about 11:00 A.M. on September 15, two waves of German airplanes (about 100 in the first wave and 150 in the second) crossed the coast near Dover, headed for London and other strategic targets nearby. A second attack of similar magnitude began at 2:00 P.M. Although largely repulsed by RAF fighters, approximately 100 German airplanes in the first attack and 70 in the second attack inflicted damage on southern and central London. Other attacks on Portland and Southampton were largely repulsed. Beginning at 8:00 P.M. and continuing throughout the night, the Germans launched wave after wave of aerial attacks.[10] It must have been frightening for the Callendars, and especially the young twins who were not yet nine years old, to live on the southern coast of England during a time of such great turmoil and to witness such an air battle. Bridget recalls seeing the blue sky that day with planes very high in the air and cartridge casings on the ground. She remembers being very confused.

Air raids continued over London and southeast England into May 1941, with the German "Blitz" inflicting considerable damage and casualties.[11] Bridget clearly remembers the blackouts, with shades and dark curtains on the windows, night patrols, and searchlights beaming across the sky. She also recalls her father, a combustion engineer, serving as a wartime "fire watcher," who, along with his neighbors, volunteered to respond to unexploded bombs or fires in the local community. Although Callendar left no record of this activity, the following personal account from Frank Mee's "My Father the Firewatcher" is typical of the rural southern England experience.[12]

> Father like all the other men in the village did extra security jobs such as Air Raid Wardens, Special Police, Fire Watchers, Auxiliary Firemen and Home Guards . . . when the sirens went and we headed for the air raid shelter he had built . . . he grabbed his stirrup pump, bucket, shovel, and tin hat heading for the church tower or some other high place where he could watch for incendiary bombs.

The reminiscences of a boy who was seven years old when the war started help to invoke the kind of world the Callendar twins must have encountered:[13]

> It didn't mean much to a seven year old until things began to happen. Curved corrugated sheets were delivered to every back garden and all the neighbours would help each other to dig big holes and use the sheets to make Anderson Shelters. The spare dirt would be thrown over the top of them and then we had a bomb-proof bedroom to keep us safe during the raids. Next to be collected were gas-masks. These were contained in cardboard boxes with a string to slip over the shoulder. Alvar Liddell said we had to carry them at all times. Gas masks had a snout on them like a pig and every once in a while new filters would be added and our snouts became bigger and bigger—I thought about Pinocchio. Babies had 'Mickey Mouse' gas masks which covered them all over and they did look a bit like their name. . . .
>
> [During the air raids] two searchlights crossed the sky—one for measuring cloud height and a larger one to seek out German planes when they nightly

droned to Liverpool with their heavy loads. Anyone who ventured out in the blackout would see a grand firework display in the night sky. . . .

Food was rationed and shopkeepers would snip bits out of your ration book when anything was bought. One of our main dishes was rabbit, and in the early morning you would see poachers returning from their night's work with red and white bellied rabbits dangling from any available bar on their bikes. The colours meant that the poor little animals had been clearly gutted on the killing fields. Rabbits were only available when there was an 'r' in the month because the rest of the time they were breeding. . . . Many houses . . . would rear one pig and would feed it on peelings and any waste food from the neighbours.

Every area had a Home Guard who were given an army uniform and an Auxiliary Fire Service who were given a boiler suit with wellies and a helmet. . . . They would practice every Sunday, hand pumping water from a brook in the field. . . . The Home Guard would parade every weekend and proudly marched around the village in their new uniforms, toting their imitation rifles. . . .

Most hardships were felt by mothers as they would queue for hours for anything going. They would knit and sew patches on holey trousers. [My mother] could reverse worn collars on shirts and do all sorts of things to make our clothes last.

Horsham, 1942

In 1942 Guy Callendar was assigned to defense work at Langhurst,[14] a secret research facility located in a country mansion in West Sussex. The well-land-scaped facility with its leafy exterior, gardens, and tennis courts blended perfectly into the English countryside and thus remained hidden from enemy attack. Here under the auspices of the Ministry of Supply and the Petroleum Warfare Department, Callendar worked on fog dispersal at airfields and other combustion projects (see Chapter 4).

The Callendar family moved into a cozy bungalow just two miles from Langhurst, at 44 Parsonage Road in Horsham where Guy resided for the rest of his life (Figure 2.7). The town of Horsham, situated halfway between London and the coast, was rural with fertile agricultural areas.[15] It was typical of British country homes to be named, and theirs was originally called "Per-cuil." In making it their own and in honor of their gardening interests, the family renamed their home "The Redwoods," and Guy retained the name

Fig. 2.7. Guy in the front garden at his home in Horsham, 1953.

Percuil for his weather station (Figure 2.8).

Guy was a peaceable, quiet, and gentle man, a seasoned and talented engineer who applied himself daily to solving the technical puzzles posed by the war. He much preferred attacking difficult engineering problems that promised to save lives rather than developing weapons that would take them. In this he followed a pattern established by his earlier experiences in World War I, including working on medical X-rays, efficiency of aircraft engines, and underwater detection devices. Because his work was classified as "top secret," Guy could not discuss the details with anyone outside the lab, including his family. This was true even on the day Langhurst called to inform Phyllis that there had been an accident at the lab and her husband had sustained an extensive but not disfiguring burn on the side of his face.[16]

The Redwoods, about one mile from Horsham town center, was very countrified when the Callendar family lived there. It boasted fruit trees, beautiful flower gardens, mature plantings, and a summer house (or shed) with windows lined with rows and rows of pots. Their apple trees produced fruit in such abundance that the family would leave baskets of apples outside the gate for others to take. Cows grazed in the field behind the house and chickens ran free in the neighborhood.

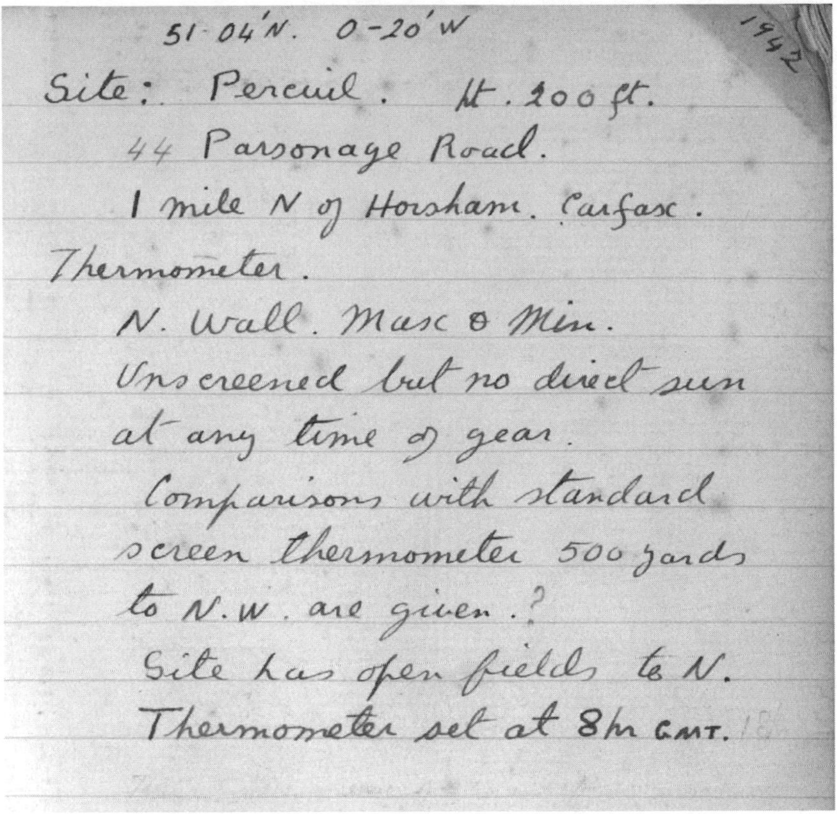
51·04'N. 0–20'W
1942
Site: Percuil. Ht. 200 ft.
44 Parsonage Road.
1 mile N of Horsham. Carfax.
Thermometer.
N. Wall. Max & Min.
Unscreened but no direct sun
at any time of year.
Comparisons with standard
screen thermometer 500 yards
to N.W. are given.?
Site has open fields to N.
Thermometer set at 8hr GMT.

Fig. 2.8. Callendar's 1942 weather diary, giving latitude, longitude, height, and details of his home weather station. "Site: Percuil, 51 04' N 0 –20' W, height 200 ft. 44 Parsonage Road, one mile north of Horsham Carfax [a historic square at the meeting place of five roads]. Thermometer. N. wall Max and Min. Unscreened but no direct sun at any time of year. Comparisons with standard screen thermometer 500 yards to N.W. are given. Site has open fields to N. Thermometer set at 8 hr GMT." From CP 2, Notebook 1942, 1.

The Callendars were fortunate to have moved to the relatively rural setting of Horsham where they could raise food and trade with local farmers, since the Ministry of Food began rationing in January 1940. A weekly ration consisted of four ounces of bacon and/or ham, six ounces of butter and/or margarine, two ounces of tea, eight ounces of sugar, two ounces of cooking fats, and meat valued to 9 pence. By 1942 domestic soap was being rationed (4 ounces of household soap or 2 ounces of toilet soap per month) as were sweets (12 ounces a month).[17] Domestic supplies reached their nadir by Christmas 1943. Wartime presents were typically practical, emphasizing gardening or canning supplies, vegetable seeds, soap, and homemade toys and crafts. The Callendar family tended a vegetable plot to provide extra food, with the girls (age 7 at the start of the war and 13 at war's end) helping out. This effort was much needed during the hard times of food shortages and strict rationing. Bridget most vividly remembers shortages in clothing, butter, marmalade, fruit, and meat, and recalls some wartime food substitutes: spam, dried eggs, and dried potatoes.

Callendar spent his evenings and weekends far removed from the war, doing private research and work at home in his small study, formerly the twins' bedroom, which became available when they built an extension onto the house. He would take temperatures for his daily weather diary—he told Phyllis and the twins he was working with his "figs." It was purely a weather journal, with no other content. For example, even on D-Day, June 6, 1944, the only entry in his diary is: "Low 47, High 65, Winds WNW at 3, 8/10 overcast, and bright early." The monthly summary for June reports, "A very windy and rather cool June. . . . There were only 2 or 3 really pleasant days and it was one of the worst Junes since the 1920s."[18]

From their Redwoods home Phyllis and the girls could easily bicycle to the market in Horsham. Bridget recalls that her mother was a "very good cook" and prepared traditional English meals for the family. For recreation they played tennis or enjoyed the public swimming pool in Horsham Park and visited the Warnham Nature Reserve nearby.[19] Phyllis took the twins to the theater every week in Horsham to see productions by such notable playwrights as J. M. Barrie and George Bernard Shaw.[20] Guy did not frequent the theater with them, but preferred to stay at home to pursue his research on climate change and manage his correspondence (see Chapter 5). Guy chose to spend his leisure time playing tennis at the Horsham Lawn Tennis Club, tending the garden, or walking and cycling in the countryside. The Callendars were pet lovers, owning a Scotty dog when the girls were young and a Dachshund named "Timmy" who joined the family in 1955 (Figure 2.11).[21]

Later Years

Gardening, tennis, bicycling, and his climate "figs." filled Callendar's later years. Guy and Phyllis would often visit Leondardslee in Sussex, one of England's most spectacular gardens, famous for its rhododendrons, azaleas, and natural woodlands and lakes. Another favorite spot was Sheffield Park, designed in the 18th century by the noted landscape gardener Lancelot "Capability" Brown. They loved the vibrant colors and intoxicating smell of azaleas (Guy's favorite flower) and planted their own at The Redwoods. Callendar also raised all his conifers from seedlings gathered from cones collected on visits to arboretums. One of the family photos shows Guy admiring

Fig. 2.9. Guy with tennis partner J. Clark, winners of the men's doubles final at the Horsham Lawn Tennis Club, 1947.

a giant sequoia seedling, *S. Giganetum*.[22] Other photos depict him in 1953 in a section of the property called "spruce valley," and in 1961 sitting contentedly among his conifers. The twins shared the family's love of gardening.

Guy also continued to stay active playing tennis and cycling. In 1947 he won the men's doubles tournament. Even in 1962, Guy was still enthusiastic

Fig. 2.10. Guy with new Phillips bicycle, 1951, having just been for a "spin."

Fig. 2.11. Guy and Phyllis in the garden, with dachshund "Timmy," 1960.

about tennis and received for his birthday "a smart new tennis bag from ma [and] balls from Bid [Bridget], so I'm well set up for tennis later on."[23]

Callendar bicycled to and from work at Langhurst each day (see Chapter 4). Occasionally he would take the family to parties at Langhurst (there were only a few) to mingle with his co-workers and their families and participate in the tennis tournaments (see Figure 4.4). He took longer bicycle trips around southern England and out "Guildford way" in the summer when the weather was permitting, a distance of 22 miles.[24] Worthing and the coast lay only 20 miles south.

Guy retired in 1958, having given "nearly sixteen years in the public service" to the Ministry of Supply and at least another 12 to his research on steam. His retirement pension, which was to begin April 1, 1958, was £293, 17 shillings, 1 pence per annum. He was also granted an additional lump sum of £662, 4 shillings, 6 pence.[25]

Guy remained very close to his daughters and exchanged letters with them, commenting frequently on their scientific interests and on the weather. In one thank you note to Anne for his birthday gifts Guy concluded, "weather is not very cold, been just above freezing for several days, but very dull and misty."[26]

His confidence in the theory of climate warming, however, was shaken by the downturn in global temperature in the 1950s and 1960s (see Chapter 5). Symbolic of this is the photo of "Dad digging us out after the blizzard, 1962."

In his monthly summary for the month of December 1962 he wrote: "a cold month notable for . . . the heavy snowfalls of the last week. Snow of 11 inches 26/27th was locally greatest for at least 20 years and total of 17 inches for last week also."[27] His weather diary for December 26 and 27, 1962, reads as follows:

Date	Min	Max	Winds	Clouds	Weather
26	22	32	SW 1	7	Fair to Snow 9"
27	25	34	NE 2	10	Snow total 11"

Callendar's family had a history of heart problems. Although he was in declining health for the last six months of his life, he did not discuss it openly. His weather journal ended in September 1964, with one entry in an almost empty grid for October. His last days were sunny ones as his journal entry

Fig. 2.12. Digging out after the blizzard, 1962.

for 1964 indicates: "The sunniest September since 1911."[28] He died on October 3, 1964, of coronary thrombosis.[29] A memorial plot for the family is located at the Sussex-Surrey Crematorium, near Crawley.

Conclusion

The first two chapters have portrayed Callendar's upbringing, early influences, scientific training, family life, and personal interests. This was done with limited documents, but thankfully, with abundant photographic resources supplied by his daughter. The picture that emerges—of a well-trained, extremely competent, pensive, and somewhat reclusive engineer, a loving husband and devoted father, and an avid sportsman—is reinforced in the next three chapters, which address his life's work and technical interests—the first of which was steam.

CHAPTER THREE

Steam Engineering

Pushing the accuracy of the measurements "as far as possible with the apparatus available"

—Callendar and Egerton

Guy Stewart Callendar became one of Britain's premier steam engineers and thermodynamicists, having learned his trade as a research assistant under his father's tutelage at Imperial College (see Chapter 1). His introduction to the world technical stage came in July 1929 when he participated in the First International Steam Table Conference held in London under the sponsorship of the British Electrical and Allied Industries Research Association (BEAIRA).[1] The conference convened engineers and physicists from Great Britain, Germany, the United States, and Czechoslovakia interested in the determination of the properties of steam over a wide range of temperatures and pressures. On Monday, July 8, Dr. Samuel Adamson, President of Institution of Mechanical Engineers (IMechE) opened the conference in the Council Chamber of the Institution. The first two and a half days were dedicated to presentations, discussions, and critiques. For the remainder of the week, delegates attempted to create standardized units and procedures, "so as to avoid in the future the discrepancies which up to the present exist between the Steam Tables used most generally by the different engineering countries of the world."[2] The conference adopted, as the recommended unit for the measurement of the total heat of steam, the International Calorie, defined as follows: "One international kilowatt-hour equals 860 international kilocalories."[3] This unit was independent of secondary properties derived from the behavior of water, local variations in the acceleration of gravity, and the value of Joule's mechanical equivalent of heat. Yet, since important theoretical questions and national engineering practices remained at issue, the delegates agreed to meet again for a second conference in 1930.

Following his father's death in 1930, G. S. Callendar rededicated himself to steam engineering. He lectured on the results of his father's research at Manchester and Cardiff[4] and the BEAIRA Turbine Committee continued funding his steam research, increasing his salary from £300 to £450.[5] Guy Callendar's work on steam was both a labor of love—an expression of filial piety for his father—and a testimony to the rigorous and practical training he had received from him.

Callendar's colleague in these efforts was the celebrated physical chemist Alfred Egerton, then at Oxford University. Funded by BEAIRA, and pursuing the agenda established by the International Steam Table Conference, Callendar and Egerton led the British effort to define international units, coordinate investigations on steam, and reduce errors and inconsistencies in its thermodynamic properties as measured by different techniques.

Alfred Egerton

Alfred Charles Glyn Egerton (1886–1959) was known from childhood to his close friends and associates as "Jack." He was educated at Eton College, where he established their scientific society, and at University College London, where he studied chemical thermodynamics with Sir William Ramsey, graduating in 1908 with first-class honors. Egerton continued his chemical studies at Nancy University (1909) and Berlin University (1913). He worked as a chemistry instructor at the Royal Military Academy, Woolwich, from 1909 to 1913. During World War I he served as an assistant in the Department of Explosives Supply, Ministry of Munitions. Here he would have likely crossed paths with Professor H. L. Callendar. In 1912 he married the Honorable Ruth Cripps, who shared his lifelong passion for travel and painting, especially watercolors. The childless couple adopted Jack's nephew Francis, whose father had been killed in action in France.

After the war, Egerton matriculated at Oxford University, where he worked as a lecturer, earning his master's degree in chemistry in 1921 and gaining promotion to reader in thermodynamics in 1923. He was elected a Fellow of the Royal Society in 1926, serving on their council from 1931 to 1933, about the time of his first collaboration with Guy Callendar. Egerton served as professor of chemical technology at Imperial College from 1936 to 1952 and as Physical Secretary of the Royal Society from 1938 to 1948. During World War II he was a member of the War Cabinet Scientific Advisory Committee and chaired the Fuel and Propulsion Committee of the Admiralty. Here Egerton would have had oversight of the FIDO project (see Chapter 4) that involved Guy Callendar.

Egerton was knighted in 1943 and was awarded the Rumford medal of the Royal Society in 1946. After the war he served on a number of scientific advisory panels, including as the director of Salter's Institute of Industrial Chemistry and as an official scientific liaison to the United States and India. He was founding editor of the journal *Fuel* and, in addition to many other honors, received the first Egerton medal from the Combustion Institute in 1958. He died suddenly, on September 7, 1959, while on a painting holiday in France.[6]

Alfred Egerton dedicated his entire life to his scientific pursuits. He was quiet and spoke little, but kept detailed daily diaries. He enjoyed long walks and skiing. "Above all he had an immensely humorous and tolerant way of seeing things."[7] Egerton was introduced to Callendar both personally and professionally when he visited Imperial College on February 20, 1930, to inspect his steam apparatus. The meeting was facilitated by Henry Lewis Guy of BEAIRA, who hinted at the possibility of funding, and Sir Henry Tizard, Rector of Imperial College, who recommended the collaboration.[8] Egerton's initial impression was that Callendar was "lacking fire." However, after witnessing an experiment he was convinced of Callendar's technical ability and was "very satisfied with the temperature measurements and with the running of the plant." He thought the experiments were working out "satisfactorily and interestingly."[9] If Guy Callendar was influenced by his father before 1930, it is clear that his mentor became Alfred Egerton after 1930. Theirs was a lifelong collaboration.

The Callendar–Egerton Collaboration

The years between 1930 and 1933 were busy ones for Egerton and Callendar. They conducted a series of experiments on the steam apparatus at Imperial College, compared their results with those of their American colleagues, prepared for the Second International Steam Table Conference in Berlin, and published a paper in 1933.

The experimental setup of H. L. Callendar was modified to provide a continuous flow of steam at high pressures and temperatures, allowing them to make "dynamical" measurements of saturation pressures (Figure 3.1). In the experimental apparatus, pure distilled water, stored in two stoneware jars, was pumped into a boiler for removal of traces of air, from which it passed into a glass storage vessel. A triple-throw hydraulic pump drew it from the vessel and forced a steady flow into the electric boiler through 19 corrosion-resistant tubes made of monel metal (an alloy of nickel, copper, and other metals). The mixture of water and steam, at pressures as high as 5000 psi, passed through the coil of the gas-heated superheater before entering the high pressure pocket. It was here that the temperature of the steam was measured with the leads of a platinum resistance thermometer, and the pressure of the steam was measured by a large differential dead-weight gauge. After passing through the throttle tube, the steam was expanded and condensed, providing a measure of its saturation.[10]

After pushing the accuracy of the measurements "as far as possible with the apparatus available," the duo found "no appreciable differences" between their technique and the American technique of making "statical" measure-

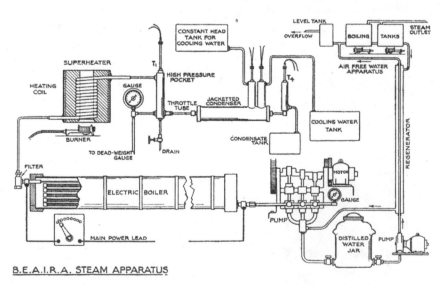

B.E.A.I.R.A. STEAM APPARATUS

Fig. 3.1. Schematic of experimental apparatus used by Callendar and Egerton for measuring the dynamical saturation pressure of steam.

ments of steam, pioneered by Frederick G. Keyes.[11] Callendar and Egerton brought their early results to the Second International Steam Table Conference, held in Berlin June 23–26, 1930.[12] Their trip was supported by BEAIRA. Although the Conference came to no consensus, and the British and American delegations disagreed on both theoretical and practical issues, all agreed that further meetings were desirable.[13]

The Third International Steam Table Conference

In 1934 Guy Stewart Callendar attended the third International Steam Table Conference held in Washington, Boston, and New York. Sponsored by the American Society of Mechanical Engineers (ASME), the conference was an international attempt to standardize the actual and theoretical behavior of steam at high temperatures and pressures. Not only was it the final steam conference in the series, it was the best documented, and arguably, the one to which Guy made the most significant contribution. Guy's love letters to Phyllis (the only ones we have), Alfred and Lady Egerton's personal journals, the conference program, and a published meeting report provide a relatively complete account of the meeting and a window into Guy's professional and personal life.[14]

Callendar's letters home are private, thoughtful, and loving (see Appendix B). While he mentioned the steam conference, he was more concerned with describing to Phyllis luxuries of the trip such as extravagant meals, opportunities to play tennis, cinematic entertainment, and personal reflections on the United States. He also frequently commented about the weather, which is no surprise due to Callendar's growing interest in meteorology.

Egerton's journal was written retrospectively and includes impressions of New York and Boston, race relations in the United States, and a visit with Henry Ford.[15] Lady Egerton's journal contains considerably more detail, including their departure by train from London with Callendar, who she referred to as "Jack's assistant and colleague on the work," and their departure on September 8 from Liverpool on the White Star Liner *Britannic*, "a very comfortable 'cabin class' oil motor ship with every luxury." Also traveling with them to the conference were H. L. Guy, "a great engineer from Manchester and his wife," and Mr. Robinson, from the BEAIRA, and his wife. BEAIRA allocated £250 for the Egerton's expenses, and probably slightly less for Callendar.

Journey across the Atlantic

The *Britannic* was a new diesel-powered ship of the White Star Line, with a low profile, sleek superstructure, and Art Deco design.[16] The captain must have been especially proud to show off the new state-of-the-art engine room to the prestigious group of engineers on the voyage. Callendar wrote to his wife, "Went over the engines of this ship with Mr. Guy yesterday, they are I.c. [internal combustion] that is like those in a motorcar. The engine room is like 500 large lorry engines all going together in a tin shed, it is no place for ladies."[17] Since diesel power left the engine room very cool, it was equipped with heaters. This was quite different from the sweltering environment of coal-fired, steam-powered engines. Callendar must have been impressed that the *Britannic*'s fuel consumption was only 40 tons per day, down 50 percent compared to steam.[18] According to Callendar, deck tennis was the most amusing game on board the ship, "it is just like Badminton except that you catch a ring instead of hitting a shuttle."[19]

As the voyagers neared the U.S. coast in the early morning of Sunday, September 16, a thick marine fog greeted the travelers as the *Britannic* docked at the Cunard Pier in East Boston.[20] They passed emigration and customs in two hours, were met by members of the American delegation, and lunched at

Fig. 3.2. Guy Callendar, 1934, "in Boston standing amongst cars with pipe and hat."

the Harvard Faculty Club (Ruth called it the "Fraternity House") hosted by Harvard professor Lionel Marks. The group toured the Harvard Museum of Natural History, motored to Lexington (Figure 3.2),[21] "where the primitive men sprang up from the farms and beat the British!", and rushed back to Boston for a bite at the Union Oyster House ("America's oldest restaurant").[22] Guy noted that Boston had nice residential districts and a magnificent harbor with small islands in it. That evening, after a very full day, Callendar, Egerton, and the other steam engineers caught an overnight train ("rather uncomfortable sleeper with common wash place") to Washington ("a fine city with solid looking white buildings").[23] Ruth remained behind in Boston at the Parker House.

Session One: Washington
After a long night of travel and breakfast on the train, the delegates arrived in Washington, D.C. and were escorted to the Broadmoor Hotel—but not to spend the night. At 10:00 AM the delegates congregated at the National Bureau of Standards (NBS) and heard welcoming addresses from its director Dr. Lyman J. Briggs and other dignitaries. After an exchange of reports and assignment of the working subcommittee (of which GSC was a part), the delegates visited the Heat and Power Laboratory at NBS and were briefed on the construction and operation of apparatus used in research on the thermal properties of steam. The delegates took a group photo (Figure 3.3),[24] lunched at the NBS, toured local points of interest, and concluded a long day with dinner at the Broadmoor Hotel.

At 9:00 P.M. the delegates reboarded the Federal Express en route to Boston for the second session of the conference—another night in transit.

Session Two: Boston
Arriving in Back Bay Station, Boston at 7:45 A.M. on Tuesday, September 18, Callendar, Egerton, and the other delegates were transported to MIT where they took breakfast at Walker Memorial and assembled in Eastman Laboratory, which Ruth called "an enormous kind of National Physical Laboratory." Harvard Professor Harvey N. Davis presided and Institute President Karl T. Compton offered formal greetings.[25] Then the delegates inspected apparatus for investigating the thermodynamic scale of temperature. Afternoon

Fig. 3.3. Participants in the Third International Steam Table Conference at the National Bureau of Standards, 1934. *Left to right, standing:* C. G. Worthington, F. Michel, E. Schmidt, G. A. Orrok, F. M. Feiker, C. B. Le Page, E. F. Mueller, E. J. M. Honigmann, J. H. Keenan, D. C. Ginnings, E. F. Fiock, H. F. Stimson. *Seated:* H. Hausen, W. Koch, H. L. Guy, A. Dow, F. Henning, L. J. Briggs, I. V. Robinson, A. C. G. Egerton, F. G. Keyes, G. S. Callendar, N. S. Osborne, H. C. Dickenson.

sessions were followed by tours around MIT and Harvard and tea in the Eastman Lab.

At 5:00 P.M. the party departed for New York City aboard the Hamburg-American Liner *New York*.[26] En route, the delegates enjoyed the sights along the New England coast and dined on board. The evening concluded with informal discussions, music, dancing, and another, more comfortable night's rest—but still in transit.

Session Three: New York

On Wednesday morning, September 19, the delegates arrived in New York City. Ruth had "never seen a more wonderful sight" as she "watched the great sky line of the sky scrapers coming nearer and nearer as we came up the river. The Empire [State] Building, the Chrysler Building, and Radio City stood out. All these looked like huge cathedral towers in pinkish white and pale blue shadows with a soft haze and gorgeous sun."[27] As the delegates settled in to the "old fashioned" Astor Hotel in Times Square, Ruth recounts meeting a number of participants including the German delegation: "Dr. Schmidt (rather a Nazi), Prof. Henning, Prof. Dreuser and nice fat Dr. Koch of Munich. They had left their wives who mostly seemed obeying Hitler's commands to increase the *vaterland*. Also a nice Austrian Dr. 'Honymann.'"[28]

The third session of the conference began at 10:30 A.M. at the headquarters of ASME in the Engineering Societies Building on 39th Street. Dr. Davis again presided and Dr. Calvin W. Rice, Secretary of ASME, greeted the delegates. That evening's formal conference dinner was held at the Hotel Astor, with Alex Dow of Detroit Edison presiding and after dinner remarks by Col. Paul Doty, President of ASME. Guy wrote that the dinner lasted from 7:00 to 11:30 P.M., and the "speeches nearly sent everyone to sleep."[29]

The remainder of the week was dedicated to the work of the subcommittee formed in Washington, "considering contributions of new experimental evidence and preparing a report for the revision and extension of the current international skeleton table."[30] Final presentations were scheduled for Saturday morning, September 22.[31] Although Callendar and Egerton had to fight for their experimental figures on total heat, they concluded in the end that the final results would be very near the agreed-upon figures. In the end the conference was voted a "tremendous success by all concerned," although, as in earlier conferences, only limited progress had been made toward a complete set of data upon which to construct steam tables, and differences in theory, techniques, and equipment prevented the publication of an international set of steam tables.[32]

Guy's experience in New York was not limited to the conference. His evenings were free and filled with various forms of entertainment. He visited the top of the Empire State Building and "looked down on all the lights of New York." He dined with the other delegates at Hotel Astor, hosted by the ASME, and enjoyed another meal at the Waldorf-Astoria, a "new, and very super hotel" with appealing "chromium plate and polished wood interiors." Their party also dined at the New York Society of Arts and enjoyed several sightseeing expeditions. After the commotion of the conference died down, however, homesickness set in:

> Now that all the rush and work of the conference is over I feel frightfully homesick, and long to be back at our quiet little home where I really belong. Once the boat starts moving I shall be all right. I see the *Britannic* started back this morning, I wish I was on her. Many people would think I must be mad to wish to quit living like a lord, free of charge, in the center of New York, but I want to get home to you and the twins. The air here is like warm thick soup, there is no movement, and it tastes & smells *very* secondhand. I have a huge fan in my room, but the air in the streets (canyons) is awful. I long for the glorious fresh breezes of Worthing.[33]

Guy spent the balance of his free time in and around New York City. He played tennis in Hoboken, New Jersey, with Professor J. H. Keenan[34] of the Stevens Institute of Technology and his friends, and was shocked to see students playing tennis with only "the shortest of shorts on, and no shirt which we should think hardly decent as there are plenty of girls about." He had hoped to travel to Philadelphia midweek to meet the Egertons, but these plans were interrupted when it was reported that the *Laconia*, the ship he planned to take home on Friday, had collided with a cargo vessel in the fog and would be out for repairs.[35] Instead Callendar sailed for home on the Cunard–White Star *Mauretania* on Wednesday, September 26, and arrived in Southampton on Tuesday, October 2.[36] It is safe to say that Callendar was happy to be reunited with Phyllis and the twins, but he also enjoyed and profited from his experiences at the Third International Steam Table

Conference, where he again rubbed shoulders and exchanged ideas with the technical elite of America and the world.

Steam Research to 1941

Callendar and Egerton's collaboration continued long beyond the steam conferences. Guy continued steam research with Edgerton for the turbine manufacturers from 1930 to 1941. Under pressure to maintain the "prestige of this country" and compete with the Americans in the race to compile steam tables, Callendar and Egerton published *The 1939 Callendar Steam Tables*, with subsequent editions, along with heat–entropy diagrams, appearing in 1944, 1949, and 1957.[37]

In 1939 there was some discussion of holding a fourth International Steam Table Conference in Prague in late June. Yet reaching international agreement on steam was an elusive goal, for as Egerton wrote, "I do not think the Conference could possibly come to agreement on [the editing of a single International Steam Table] at present . . . [since it] would need to be thermodynamically consistent throughout and to lie within the agreed errors of the observed results."[38] Regardless, the conference was never held due to the threatening political situation in Czechoslovakia.

By September 1939, with the outbreak of war in Europe, Egerton and Callendar had to make emergency contingency plans for the fate of their steam apparatus at Imperial College. Options included (a) continuing the work on site, (b) leaving the apparatus intact during the war, and (c) transferring the apparatus to a "neutral" zone. Although concerns were raised about the safety of London, the priority of steam research, the cost of moving the apparatus, and possible personnel shortages, we know that the apparatus remained intact through 1941. Early in the war, Callendar busied himself with compiling a concise final report from all the earlier work that had been completed.[39] He continued his steam research, attending BEAIRA subcommittee meetings and reporting on his attempts to measure the total heat of steam at higher temperatures and pressures.[40] However, at this point it was clear to all parties involved that the work of Egerton and Callendar on steam research was coming to an end.

The steam work came to a formal close at a BEAIRA Turbine Research Committee meeting on July 11, 1941. Callendar, present by invitation, commented on the investigations and asked that the apparatus be kept set up at Imperial College in case further testing was needed. It was noted that Professor Egerton was preparing a final report for publication, and the Committee recorded its appreciation of the work done by Egerton and Callendar.[41]

"What is to become of Mr. Callendar?"

On August 5, 1940, with the Battle of Britain raging, H. L. Guy of Metropolitan Vickers wrote to Dr. E. B. Wedmore, the director and secretary of BEAIRA, about the steam table work nearing completion and the question of "What is to become of Mr. Callendar?" Since Callendar had worked on the steam tables full time for BEAIRA, "ever since he was put on to it at the suggestion of his father," and was considered to be a "very competent experimentalist," Mr. Guy wondered if Callendar could be incorporated into the staff of the BEAIRA Laboratories when his steam work ended.[42] This letter launched a series of consultations and inquiries aimed at finding Callendar a job. Wedmore responded that he too had been "a little troubled as to Mr. Callendar's future," and welcomed Mr. Guy's cooperation. However, he was not very hopeful that Callendar would fit into the work being done at the BEAIRA Laboratory, which was "rather highly specialized" and focused on electrical surge phenomena, radio interference, and the properties of dielectrics. Wedmore thought that Callendar might find a more suitable position in industry or perhaps with the National Physical Laboratory and suggested they have a "frank conversation" with him about his future in which he would "no doubt discuss what he must already have on his own mind."[43] Egerton's correspondence has the annotation "saw Callendar" and includes a note about a possible grant for investigating the thermal properties of hydrocarbons that could provide £500 per year.[44] H. L. Guy was of the opinion that "we should do our best for Callendar," and suggested to Wedmore that Sir Henry Tizard be involved in the discussions.[45]

Egerton was well positioned to assist in Callender's job search. He was chairman of Chemical Technology at Imperial College and Physical Secretary of the Royal Society. In addition to steam, he conducted research on fuels and combustion. During the war, he served as a member (one of only six) of the War Cabinet Scientific Advisory Committee and chairman of the Fuel and Propulsion Committee of the Admiralty.[46] Callendar's résumé of 1940 was in Egerton's possession and is reproduced in Figure 3.4 below.[47]

MR G. S. CALLENDAR

Born 1898. Educated at St. Paul's School, and City and Guilds Engineering College.

During the last war assisted in testing a variety of generating sets for X-ray apparatus, and aircraft engines at Farnborough. Later joined R.N.V.R. as hydrophones officer and had experience with electrical apparatus used for sound ranging.

Assisted Father, the late Professor H. L. Callendar, in numerous researches mainly of an electrical or thermal nature, including his tests on turbines at various power stations.

In collaboration with Father and Professor Egerton designed and operated the E.R.A. steam research apparatus at South Kensington. In this connection have had considerable experience in making out reports and tables dealing with the researches. Have published a paper on electrical thermometry in the Philosophical Transactions of the Physical Society.

Is a Fellow of the Royal Meteorological Society, who has published papers by him dealing with radiation from gases in the atmosphere.

At the invitation of the Air Ministry wrote an article on the past history of the atmosphere, which they published in full in the Meteorological Magazine.

Have knowledge of the use and testing of fuel: in this connection have been invited to attend the visitors day at H.M. Fuel Research Station.

In general his qualification is over 20 years continuous experience in precision measurements, and especially in detecting small errors and designing apparatus to overcome them.

References to papers

The numerous reports on various aspects of the steam research, which have been issued by the E.R.A. (Refs. J/T and JCT) are of course familiar to you.

Besides the above the principal papers under his name are as follow:

"On the vapour pressure of water from 150° to the Critical point." (With Prof. Egerton) Phil. Trans. Roy. Soc. 1931.

"On the reduction of Platinum Resistance thermometers to the International Temperature scale." Phil. Mag. May 14. 1932.

"On the influence of Carbon Dioxide on Earth temperatures." Q.J. Roy. Met. Soc. 64. 1938.

"The atmosphere through the ages." Met. Mag. March. 1939.

"On the variations of Carbon Dioxide in the atmosphere." Quart. J. Roy. Met. Soc. October. 1940.

November 8th 1940.

Fig. 3.4. Callendar's résumé, 1940.

Of course, the technical demands of World War II ensured that G. S. Callendar would remain fully employed (see Chapter 4).

A Hiatus of Nineteen Years

Publication of the Callendar-Egerton steam research results was not completed until 1960. Soon after the end of BEAIRA sponsorship in 1941, Wedmore wrote to Egerton asking about the status of his final report. Egerton replied, "the trouble is that it requires freedom of mind for a few clear days and that I have not got, as soon as I get free something immediately connected with the war crops up which has to be attended to. . . . I am afraid it cannot be what I hoped it might be, for I wanted to get suitable equations that would express the results and the whole work together, but that must wait till after this war."[48]

Publication of the steam research conducted by Callendar and Egerton was in fact postponed many years beyond the war and didn't occur until 1960 in the form of Egerton's final posthumous publication, together with Callendar, in the *Philosophical Transactions of the Royal Society*.[49] In a letter to the Royal Society early in 1959, Egerton explained, "The Total Heat study was ten years of work (1930–1940), but I never had time to get it in form for publication until last summer! There is renewed interest in the subject, particularly in America and Russia, and so I felt it my duty to make the effort to condense the voluminous reports which we had sent during the progress of the work to the Electrical Research Association, and to provide a paper which would be useful to those now working in the field. . . . I hope that the publication will be agreed, in spite of the belated writing up of the work."[50] His diary for December 15, 1958, also notes that "Callendar has written approving the steam paper."[51]

CHAPTER FOUR

Defense Work

We had been making vast preparations to cook the Germans. We would see whether we could cook the atmosphere!

—Donald Banks

World War II generated a high demand in Britain for talented scientists and engineers. Callendar responded to the call of duty, as he had in World War I, and actively participated in research and development for the Petroleum Warfare Department and the Ministry of Supply. He was a key engineer and shared a patent in FIDO, an airfield fog dispersal system that saved thousands of airmen's lives and, according to most contemporary reports, shortened the war by two years. The bulk of this chapter focuses on FIDO (Fog Investigation and Dispersal Operation), as it was Callendar's greatest wartime achievement. However, it was not his only accomplishment during World War II. He conducted basic and applied research on the efficiency of fuel cells and on the infrared spectrum, the latter resulting in strategically important knowledge about German fuel supplies, as well as yielding fundamental insights into the roles of water vapor and carbon dioxide absorption and emission in the Earth's heat budget. He also designed internal baffles for fuel tanks, experimental devices for forest clearing, and new fuel propellant systems and flamethrowers. After the war, Callendar continued research for the Ministry of Supply on combustion technologies and initiated a number of new projects, including testing of space heaters for military purposes and methods for generating high-speed air currents. Throughout his years of government service—he retired in 1958—Callendar remained steadfast in his resolve to contribute peacefully, and did not choose to direct his research toward violent ends.

Beyond Steam

As research on steam was winding down, Callendar was given a temporary assignment by BEAIRA to examine different types of electrochemical storage and generating devices. This work required that Callendar devise tests to record and compare the efficiencies of different types of batteries and fuel cells.[1] The challenge was immense, since the problem of providing powerful, reliable, and long-lived sources of electricity in the field was a perennial one and, to date, no fuel cell had yet been operated successfully.[2]

Callendar's atmospheric interests were also relevant to the war effort. Early in 1942 Callendar collaborated with Gordon Brims Black McIvor Sutherland, a Cambridge physicist who had taken on the task of using infrared spectroscopy to analyze the aircraft fuel being used by the Germans. Under sponsorship of the Ministry of Aircraft Production, Sutherland and his assistants examined samples of captured German fuel and measured the percentage absorption in the infrared spectrum. According to Sutherland's biographer, the "complex vibration-rotation bands from atmospheric CO_2 and H_2O in the spectrometer optical path caused difficulties in measurements."[3]

At this point, Sutherland turned to G. S. Callendar, who had recently published a paper in the *Quarterly Journal of the Royal Meteorological Society* on infrared absorption by atmospheric CO_2 (see Chapter 5).[4] In this paper, Callendar reviewed laboratory observations of the mean absorption by CO_2 at standard temperature and pressure and presented an empirical formula that could be applied to environmental conditions. He also discussed other factors such as the effect of pressure on absorption, water vapor overlap, and the overall heat balance of the atmosphere. Working with Sutherland and his team, Callendar was able to investigate the atmospheric conditions that were complicating the analysis of German fuel samples. This work was classified.

The Gassiot Committee of the Royal Society, which specialized in upper air research, was particularly interested in radiation equilibrium conditions in the stratosphere. Under the chairmanship of G. M. B. Dobson (Egerton was also a member), they commissioned a study by Sutherland and Callendar on the infrared spectra of "all the possible atmospheric constituents," with special reference to H_2O, CO_2, and O_3. The paper discusses the presence of major discrete absorption bands based on the vibration-rotation spectra of these three molecules and calls for more experimental measurements to investigate Doppler broadening, collision broadening, and the appearance of additional lines and bands at higher temperatures. Overlap of the absorption bands of each trace gas with varying amounts of water vapor was of interest, as was the need for new accurate measurements of the absolute absorption coefficients of a variety of molecules: NH_3, CH_4, C_2H_4, C_2H_6, C_3H_8, C_2N_2, HCN, N_2O_4, N_2O_5.[5] Of course, these measurements had to be done in complete secrecy and only a portion of the results appeared in the open literature.[6]

Langhurst

In June 1942 Callendar was officially assigned to the Ministry of Supply and was transferred to the research staff of the newly organized Armament Research Department (ARD) at Langhurst, Sussex.[7] The Langhurst team had recently been assigned to a high priority mission directed by the Petroleum Warfare Department (PWD), an agency created in 1940 to consider "the possibilities inherent in the use of burning oil as an offensive and defensive weapon in warfare."[8] Other PWD projects included operation PLUTO,[9] "Pipe Line Under the Ocean," a thousand-mile underwater cross-Channel petrol pipeline; land and sea flame barrages protecting the British coastline;[10] and flame-throwing tanks like the Churchill "Crocodile" and the "Wasp."[11]

Under the leadership of Britain's Minister of Fuel and Power, Sir Geoffrey Lloyd, and Major-General Sir Donald Banks, the PWD supported experimental work on flamethrowers at its Proofing and Experimental Establishment at Langhurst.[12] It also supervised the development of an operational airfield fog dispersal system, FIDO. Immediately upon being assigned to Langhurst, Callendar began work on this top-secret project.

FIDO and the Problem of Fog

The frequent presence of thick fogs on British airfields scuttled many planned air raids against Germany; moreover, visibility conditions could deteriorate quickly enough to endanger the safe return of exhausted aircrews, in many cases low on fuel, and with serious injuries to their personnel and structural damage to their airplanes. In 1940 alone, poor visibility conditions were the cause of 21 crashes in September, 32 in October, and 34 in November—fog had become a "venerable opponent" for Royal Air Force (RAF) pilots.[13] Since blind landing techniques were not available at the time, researchers, with support from the highest levels of government, turned their attention to fog dispersal. As Prime Minister Winston Churchill quipped when he learned of the situation, "There is no need to fight the enemy and the weather at the same time!"[14]

Under the leadership of Lord Cherwell, the War Cabinet (including Egerton), the scientific research establishment, and industry joined forces to tackle the problem. The result was the creation of FIDO, which began as a research project and developed into a reliable operational system to clear fog from airfields, allowing British and Allied aircraft to take off and land in conditions of poor visibility. The FIDO system involved a system of tanks, pipes, and burners surrounding British airfields delivering petroleum that, when ignited, raised the ambient temperature by several degrees—enough to disperse fog and light the way for aircraft operations. It was a massive undertaking, but its success was in large part due to the brilliant, but modest engineer at Langhurst, G. S. Callendar, who designed key components of the system, including the trench burners, and who was one of the FIDO patent holders.[15]

Unquestionably, airfields in the British Midlands were particularly susceptible to fog formation:

> On a clear, calm autumn or winter night, the ground temperature falls considerably after sunset, chilling the air above it. . . . [The] greatest effect is felt in the few hours before dawn. Because the heat has been lost due to radiation from the earth's surface, the fog which results when the air temperature falls below its dew-point is known as radiation fog and is the commonest on British airfields.[16]

This thick radiation fog—also known as ground fog—shrouded the airfields on many nights designated for bombing raids. Other types of fog generated by larger-scale weather conditions, such as advection fog or low-level stratus clouds, also posed problems. But natural causes were not the only source of fog formation: Smoke from countless coal-fired factory and domestic chimneys provided "millions of minute particles of soot which provided nuclei for condensation drops to form as they gradually drifted out to country areas where radiation fogs were forming and where airfields were situated."[17] The combination of British weather patterns and atmospheric pollution gave rise to frequent "peasoup" fogs that endangered the safety of British aviation and the efficacy of the war effort.

Early Ideas about Fog Dissipation

During World War I, it was accepted that the presence of fog meant that aircraft would be grounded.[18] By 1921, Sir Napier Shaw, Professor of Meteorology at Imperial College, discussed the possibility of clearing fog at an aerodrome by heating it, concluding, "I would not like to say it is impossible with unlimited funds and coal." He noted, however, that "air in the open is very slippery stuff and it has all sorts of ways of evading control that are very disappointing."[19] Professor F. A. Lindemann (later Lord Cherwell, who initiated FIDO) agreed with Shaw and chose to emphasize blind landing techniques.[20] Other possibilities, though none of them proven, included sprays of electrified water, air, or sand; chemical treatments; vigorous fanning; and coating rivers with oil.[21] Yet the brute force technology of heating the runway was the only one certain to work—although it was prohibitively expensive. In 1926, W. J. Humphreys, director of research for the U.S. Weather Bureau, estimated that it would require the combustion of 6,600 gallons of oil (or 35 tons of coal) per hour to clear a 50-meter-thick layer of fog from a typical aerodrome—a cost far too large.[22]

Professor David Brunt, Shaw's successor at Imperial College, revisited the issue in 1939. He estimated that clearing a 100-meter fog layer would require an average temperature increase of 3.5°C (twice this at the ground) and suggested that smokeless burners supplied by an oil pipeline along an airfield could be designed to do the job.[23] Brunt's ideas were tested in the winter of 1938/39 at Farnborough, but the results were not promising.[24] Suggestively, Callendar (who had worked at Farnborough during the Great War) included the following line in his résumé of 1940: "Have knowledge of the use and testing of fuel: in this connection have been invited to attend the visitors day at H. M. Fuel Research Station."[25] With his connections to Imperial College,

Farnborough, and leading British scientists, Callendar was well prepared for his work on FIDO.

As the war escalated, fog became an obstacle to successful bombing raids. With more raids scheduled, a surging accident rate and the large number of flying hours lost to fog made the problem one of "extreme urgency."[26] Responding to the problem, Churchill directed his scientific advisor Lord Cherwell to address the matter and issued the following statement:

> It is of great importance to find means to disperse fog at aerodromes so that aircraft can land safely. Let full experiments to this end be put in hand by the Petroleum Warfare Department with all expedition. They should be given every support.[27]

Under the leadership of Lloyd and Banks, the Petroleum Warfare Department assembled a technical team, reviewed earlier work on the fog problem, and consulted with their colleagues in America and in British industry, quickly reaching the conclusion that heat was the best option for fog dispersal.[28] According to Banks, "We had been making vast preparations to cook the Germans. We would see whether we could cook the atmosphere!"[29]

The FIDO project brought together pilots, engineers, fuel scientists, and meteorologists.[30] Given the urgency of the situation, normal research and development plans were shelved in favor of an all-out attack by different groups including scientists and engineers from the National Physical Laboratory, Imperial College of Science and Technology, Royal Aircraft Establishment, and of course, Langhurst, and industries such as the Anglo-Iranian Oil Company Ltd., Gas Light and Coke Company, General Electric Corporation, Imperial Chemical Industries Limited, London Midland and Scottish Railway, and the Metropolitan Water Board. According to Lloyd, "each was told to get on with the job with the fullest support and freedom of action."[31]

By early 1942, Bomber Command had articulated its needs to the PWD, calling for a method of dispersing radiation fog to a height of 100 feet from a runway 50 yards wide and at least 1000 yards long, without placing any obstruction within 50 yards of the runway edges."[32] With these specifications in mind, two experimental installations were agreed upon by October 10, 1942 (about the time the Callendars moved to Horsham). The first was a petroleum-burning fog dispersal system to be installed at the RAF base of Graveley, and the second was a coke-burning system at RAF Lakenheath.[33]

First Successful Tests

The first large-scale test of a FIDO system was conducted in a field and did not involve aircraft takeoffs or landings. With strong radiation fog predicted for the morning of November 4, 1942, the FIDO team assembled at Moody Down, Hampshire. An 80-foot fire escape ladder was positioned between two FIDO burners 200 yards in length and 100 yards apart. As a local fireman climbed to the top of the ladder, he disappeared into the fog. When the burners were lit, the fog began to clear and the fireman came into view. To verify the result, the burners were turned down and the fog reappeared. The burners were again ignited and the fog dissipated. With typical British reserve, it was reported that "Geoffrey Lloyd *almost* whooped for joy."[34]

On the same day, experiments were also conducted at Staines using coke-burning braziers distributed by miniature rail cars along tracks paralleling the runways. While an even denser fog was cleared with less smoke, the coke took longer to light and required more effort to replenish. Petrol was much easier to supply to airfields and ultimately became the fuel of choice for FIDO. The urgency of the situation did not allow much time for further experimentation and research. As a result, the petroleum burner setup at Graveley served as the prototype for other FIDO systems ultimately installed at 14 RAF airfields.

FIDO assumed the same function at each of these airfields. Burners, protected by windscreens, were positioned parallel, yet distant, from the landing strip. Vaporized petrol exited through holes drilled into pipelines, creating smokeless flames that could reach 3 feet or higher. This created a vast hollow box of flame around the airfield that heated the atmosphere and cleared the fog.[35]

On February 5, 1943, Air Vice Marshal D. C. Bennet landed a Gypsy Major at Gravely in a midday FIDO light-up. Thirteen days later, in the first night test, he again landed, this time in a Lancaster. Although it was not foggy, visibility was poor. Bennet recounted seeing the runway when he was still 60 miles out. As he made his approach he recalled, "I had vague thoughts of seeing lions jumping through a hoop of flames at the circus. The glare was certainly considerable and there was some turbulence, but it was nothing to worry about."[36] Except wildfires. A demonstration test for aircrews on February 23 resulted in grass, hedges, trees, and telegraph poles near the burners going up in smoke.

The first opportunity to test land aircraft in foggy conditions occurred in July 1943. A thick fog, approximately 300–400 feet deep, blanketed the runway; visibility was less than 200 yards. The FIDO burners were lit at 5 A.M., and within seven minutes, an area 1500 yards long and 200 yards wide was cleared of fog. To the delight of Geoffrey Lloyd, aircraft landed successfully at 15-minute intervals.[37]

Callendar's Contribution

FIDO engineers confronted several significant problems in designing an operational fog dispersal system. Smokeless burners were needed that were effective, fuel efficient, and functioned in windy conditions. With flames rising as high as 3 feet in the air, it was imperative that smoke and fiery glare from FIDO itself not reduce visibility on the airfield. In the laboratory, preheating the fuel to vaporize it and optimizing the fuel delivery system could achieve a smokeless burn. But ambient winds could interfere with the burners. Early experiments revealed that a FIDO system even generated its own local winds of up to 12 miles per hour when a heated mass of air rose over the airfield and converging air rushed in to replace it. In order to function properly, the burners needed to be shielded from both the winds and their own back draft.

At Langhurst Callendar studied earlier experimental work on the distribution of radiation from exposed flames—a topic related to the efficiency of fuel burning. He concluded that a nonluminous flame is about 6 percent more efficient as a source of heat than is a luminous one.[38] He also designed a 30-foot model of a FIDO system that had "a stable performance in moderate winds." It employed preheated petrol, large bore pipes, and careful burner spacing. To reduce glare and provide additional shelter from the wind, Callendar experimented with trench burners, designing a system to be installed below ground level. This project was dropped, however, when the trench burners failed to generate sufficient heat output and otherwise failed to meet his strict efficiency criteria.[39]

Some of Callendar's experiments were conducted indoors, at Empress Hall, Earl's Court, a huge skating rink in central London where the ice-making machinery was used to produce synthetic fogs and a large wind tunnel was constructed to simulate airfield conditions.[40] Using this experimental apparatus, various arrangements of model burners could be observed in different wind conditions. One year after the conclusion of the war, the official report noted that the experiments at Empress Hall "enabled [the] experimental information to be obtained, and installations to be designed within a period far smaller than any that would have been possible had it been necessary to rely upon full scale observations in the open for all the information that was wanted."[41]

When A. C. Hartley, chief oil engineer for the Anglo-Persian Oil Company and technical director of the Petroleum Warfare Department, visited Langhurst in January 1943 he "saw small burners being developed by Mr. Callendar." One of these was ignited with butane and, when the pipes were sufficiently heated, ran on vaporized petrol. Based on what he learned from Callendar, Hartley arranged immediately for small holes to be drilled in the existing FIDO burners "so that some flames would jet downwards on to the preheater pipes and increase the vapourisation." The result was "a great improvement in performance when I saw it tested on January 23rd."[42] The final FIDO design, the so-called Haigill (Hartley Anglo-Iranian Gill) burner thus owed its creation to improvements made by Callendar. As Banks writes, "This became the accepted system and was successfully installed and operated at most of the airfields where FIDO was subsequently fitted."[43]

Callendar's list of his own technical reports for the PWD include the following nine on FIDO:

Petrol vapour discharge
A 30-foot model FIDO burner
A self-evaporating liquid fuel burner of the air induction type
On the use of leaded petrol in evaporating burners
Trench burners for closing a runway to fog intrusion
The trench burner
The distribution of radiation from exposed flames
Experiments on the thermal melting of ground ice
The de-icing radiator[44]

As in his earlier collaborations, Callendar's work on FIDO brought him into close association with high-level administrators and scientists, notably Lloyd and Banks, who visited Langhurst to see his experiments; A. O. Rankine, formerly of Imperial College; and Sir Nelson Johnson, head of the Meteoro-logical Office.[45] While most of his work was done at Langhurst, Callendar also conducted experiments in London and at FIDO airfields, and attended Petroleum Warfare Department meetings when needed.[46]

FIDO Becomes Operational

The experience of landing at a FIDO airfield was both unsettling and exhila-rating. After reviewing the fog and weather report from the Meteorological Office, the airfield administrator would order the FIDO system operational. This process took nearly 20 minutes and involved a crew of at least a dozen workers to complete. Once lit, the burners made a loud roaring noise that would drown out the sound of approaching aircraft.[47] Suddenly, a low-fly-ing aircraft would burst through the fog, in some cases only 100 yards away from the airfield and continue its approach, bouncing momentarily as it crossed the turbulent zone of the burners, and touching down safely on the runway.[48]

The view from the cockpit, however, was entirely different. Although air-men were thankful for the safety that FIDO provided, they described their first experiences of landing between FIDO burners as frightening. One vet-eran pilot likened it to a descent into hell, remarking that it seemed as if "he was over target once more . . . [and] that the whole place must have caught on fire."[49] After the novelty had worn off, aviator comments included, "Piece of cake!"; "I couldn't believe it possible—but now I know"; "It's my navigator's salvation. We saw it 150 miles away"; and "I'd do it a hundred times and like it."[50]

The urgency that Prime Minister Churchill demanded had been met, and FIDO was quickly serving the duty of guiding RAF airmen home safely. Geoffrey Williams writes:

> From the time of the first operational use of FIDO at Graveley on the 19/20 November 1943 until the end of the year, thirty-nine successful landings were made. By the spring of 1944, eight FIDO-equipped airfields were in operation, and seven more . . . were under construction.[51]

Because of FIDO, the Allies could launch patrols and air raids and return their planes safely when enemy aircraft were grounded due to poor visibility.

Fig. 4.1. Boeing B-17 Flying Fortress, 493rd Bomb Group, landing in England with aid of FIDO, November 16, 1944. U.S. National Archives Photo (A9004, detail).

RAF Coastal Command aircraft on anti–U Boat patrol used FIDO frequently. On one occasion, a lost Lysander aircraft landed on a runway that had been cleared of fog. When FIDO was turned off, the Lysander, unbeknownst to station authorities, remained on the runway and was enveloped by fog. It took the pilot two hours to find the control tower! Just before the end of the war, Berlin Express Mosquitos made raids on 36 consecutive days from fogbound East Anglia: FIDO was responsible for their regularity. Between 1943 and 1945, a total of 2,486 British and American aircraft landed when FIDO was operational; 722 of these landed with visibility conditions of less than 1000 meters; and 79 under ceilings of less than 100 meters.[52]

The success of FIDO was presented to a war-weary public as almost a miracle. Newspapers proclaimed it as a lifesaver and a triumph for British aviation and those involved in administering the project credited FIDO with shortening the war and saving the lives of over 10,000 airmen.[53] The voice of the airmen echoed this praise. Without doubt, FIDO was favored by pilots returning to foggy England after a mission, since they could see the airfield glowing in the distance, beckoning them home to a lighted, fog-

Fig. 4.2. U.S. Navy RD4 lands with aid of "FIDO," October 3, 1945, at U.S. Naval Auxiliary Air Station, Arcata, California. Picture was taken after the heating equipment had dissipated the fog sufficiently for the plane to make a normal landing. The plane was guided into the cleared area by radio. Heavy fog concealed the runway until the burners were lit and within a very few minutes after the burners were turned off heavy fog was again present. Ivan E. Anderson to Chief of the Weather Bureau, July 11, 1946. U.S. National Archives Photo (27-G-1A-8-317-O).

free airport. They could also save valuable time getting their shot-up planes and exhausted (and possibly wounded) crews on the ground. "Ninth Time Lucky," from the airman's newsletter *Tee Emm*, recounts a crew's harrowing experience in eight failed landing attempts in fog, followed by a successful, picture-perfect ninth attempt after the FIDO burners were ignited (Figure 4.3).[54]

Military historians are fond of invoking the "the fog of war" as they struggle to reconstruct events. In the case of England the fog was literal, for example in the opening days of the Battle of the Bulge, when FIDO supported the Allied aviation. But during the long campaign the weather cleared and much of the tactical air support for the campaign came from the Continent, not England. Thus contemporary evaluations of the overall success of FIDO

FIDO
NINTH TIME LUCKY
TEE EMM 1944

Fig. 4.3. "Fido is going around throwing a chest like a bull-dog." Cartoon from *Tee Emm* 4 (October 1944): 155.

in "shortening the war" may have been overly optimistic and self-serving, given the controversial role of strategic bombing in World War II, the intervention of Soviet forces on the eastern front, and the fact that the war in the Pacific was still raging.[55]

The Aftermath of FIDO

FIDO proved to be one of the innovation success stories of World War II. It was a crash research program that became operational; it saved lives and equipment; and it definitely gave the edge to Allied aviation during the last two years of the war. But FIDO was feasible only under the desperate conditions of wartime. Bomber Command, its chief beneficiary, credited it with introducing a "revolutionary change in the air war," but its success was never replicated.[56]

When the FIDO system was ignited at an airfield, *6,000 gallons* of petrol were burned during the 4-minute period it required to land one aircraft. By comparison, a Mosquito bomber might burn between 10 and 20 gallons of fuel during its landing approach. During the two and a half years it was in operation, it is estimated that FIDO consumed a total of 30 million gallons of petrol.[57] Such expenditures were justifiable only when national survival was at stake.

After the war a FIDO system was planned for London's Heathrow Airport, but it was never installed. For a time the FIDO systems were maintained at two RAF bases at Blackbushe and Manston, but according to one estimate, done in 1957, the cost of running a FIDO installation was prohibitively expensive—£44,500 per hour. Experiments using jet engines installed along runways to heat the air and disperse fog at Orly Airport near Paris and in Nanyuan, China, met with mixed results.[58] The main technique for dealing with fog, developed after the war, was not weather modification, but the widespread use of instrumented landing techniques.[59]

Callendar's Other Projects for PWD
Callendar was deeply engaged in a research program that emphasized the development of flamethrowers. Thus, it was impossible not to contribute to their use as offensive weapons. Nevertheless, by emphasizing FIDO and other projects involving the efficiency of fuel propellant systems and the application of thermal technology to war-related problems, Callendar attempted to isolate his work as much as possible from the killing fields. In addition to his work on the FIDO system, he authored the following reports for the PWD:

Hydrogen production
The production of permanent gas at high pressure from a single liquid
The use of ammonia as a coolant for cordite gas
The use of dissolved carbon dioxide in pressure accumulators
Trial of a German flamethrower
The use of steam for discharging fuel tanks of flamethrowers
The use of petrol vapour for ejecting flamethrower fuel
Tests with anti surge baffles in a tank
A vapourized petrol burner for flamethrower ignition
Pressure losses with fuel in straight pipes
Experiments on the thermal cutting of wood[60]

Several examples from this long list will suffice. With British forces in Burma facing "a formidable obstacle to military operations" in their long campaign in the tropical jungles, Callendar was assigned to investigate new technologies for clearing forests. His tests showed that, due to its high moisture content, cutting green wood with flames was inefficient, even when using an oxyacetylene blowtorch. He concluded that "the fastest flame cutting is only about one hundredth of that which could be obtained with a small mechanical saw."[61] He also tried cutting sticks with electrically heated nichrome wires, but the wires often broke during the tests. Callendar concluded that these two thermal techniques may be capable of removing small vegetation—brushes and scrub—but not large trees. Callendar also investigated the most effective type of mechanical baffles to install in large fuel tanks to prevent fuel from piling up against the side when vehicles in motion accelerated. These antisurge baffles had to be designed to allow the free flow of fuel while preventing loss of capacity and gas penetration into the tank through the discharge pipe.[62] Callendar was also credited with the development of a pyrotechnic device for destroying enemy oil tanks for use in raids by British Commandos and Special Air Services.[63]

Post-war Defense Work
After the war, Callendar continued his employment at Langhurst for nearly 12 years. When the PWD was disbanded in 1946, work on flame warfare was continued by the Ministry of Supply's Armament Development Establishment ("Research" was added in 1954). Retained copies of his research reports from the 1950s include the following:

Trial of a Self Contained Portable Space Heating Unit
The Diffusion of High Pressure Air into Liquids through Flexible
    Membranes
Gravity Method of Obtaining a Low Pressure High Velocity Air Current
    for Laboratory Research

With Britain facing possible military operations in arctic climates during the Cold War, the military needed safe and transferable units to heat military buildings and tents in cold climates. This was important both for personnel as well as for storing ordnance safely and maintaining equipment at effective operating temperatures. Callendar was asked to evaluate the "Dragon" self-contained portable space heater. Compared to a conventional stove, the Dragon heater was more expensive, but was far more efficient. Moreover, because it circulated fresh air, it could be operated in a mode that prevented the buildup of explosive or toxic vapors and prevented the pooling of hot air within a building—important considerations when storing combustibles such as fuel or munitions.[64]

Callendar's work with the diffusion of high-pressure air or other gases into liquids certainly reflects his wartime research experience with the PWD. When using high pressure gas to expel a liquid (such as petrol) from a vessel (such as a flamethrower tank) two difficulties arise: "one is the tendency for the gas to find its way to the outlet of the vessel before all the liquid is

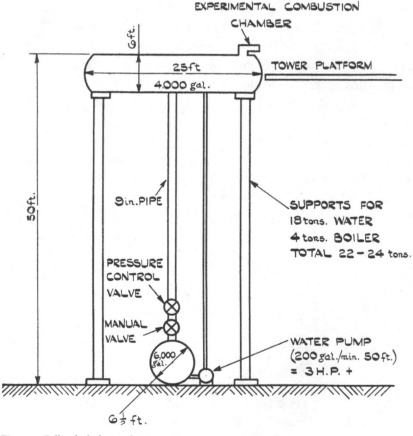

Fig. 4.4. Callendar's design for generating a supersonic jet of air.

discharged. This is called channeling." The second is the tendency, over time, for the gas to be absorbed into the liquid, resulting in potentially undesirable effects. Callendar determined that an impervious flexible membrane interposed between the gas and the liquid created the most effective pressure possible and reduced the absorption of gas by a factor of 10 or more. This allowed the liquid to remain pressurized longer without adverse effects. His trials indicated that rubber and plastic membranes worked best, and that the membrane could be protected from corrosion by a thin coating of aluminum.[65]

Callendar's design of a giant water tower for generating a supersonic jet of air was probably never built. In the diagram (Figure 4.4),[66] water released from the upper holding tank falls through a 9-inch pipe into the lower tank, causing the pressure to fall to 0.1 atmosphere. The air rushing into the tank across the small venturi orifice to accelerate the flow could approach 1200 miles per hour.[67] Combustion experiments could then be conducted in the supersonic airstream in the combustion chamber.

Callendar was a well-respected engineer at Langhurst, earning £550 per annum, the top salary of all the engineers, when he joined the research group in 1942.[68] Over the years Callendar listed his job title as variously "Senior Scientific Officer," or "Principal Scientific Officer" with the ADE.[69]

Conclusion
Using scattered reports, journal entries, government documents, and various reference materials, this chapter has only partially outlined some of the

Fig. 4.5. Guy and Phyllis at a party at Langhurst in 1955.

Fig. 4.6. Guy supervising combustion at home with his own "fido," Timmy.

important contributions made by Guy Stewart Callendar during his working years. It is difficult to detail exactly all of Callendar's projects during the very secretive years of World War II and the Cold War, as most of his work was classified or highly confidential. He could not discuss his projects at home with his family. In fact, the only exposure his family had to Langhurst was through the few parties that were hosted there (Figure 4.5).

By all accounts, FIDO was a lifesaver for the Royal Air Force in World War II, and G. S. Callendar was one of its key engineers and patent holders. With the FIDO system operational, aircraft were not only safe to land in fog, but also bombing raids on foggy nights, at one point considered impossible, were carried out in the glow of FIDO. Although FIDO may have been research-intensive and expensive, it is credited with shortening the war and saving thousands of airmen's lives. Thus it is only sensible that aviators of the RAF were dog lovers; FIDO was certainly a man's best friend (Figure 4.6).

CHAPTER FIVE

Global Warming and Anthropogenic CO$_2$

How easy it is to criticise and how difficult to produce constructive theories of climate change!

—Callendar[1]

G. S. Callendar, working largely alone and from home, established the carbon dioxide theory of climate change in its essentially modern form. He studied climate change, including global temperature trends, atmospheric concentrations of carbon dioxide, and the infrared absorption and emission spectra of trace gases. He also investigated the carbon cycle, including natural and anthropogenic sources and sinks, and the role of glaciers in the Earth's heat budget—all in an era before computer climate modeling. The data he collected on temperature records and trends worldwide filled dozens of his research notebooks; the notable series of articles on climate change he published between 1938 and 1964 were the result of his extensive calculations, laboratory measurements, literature searches, and professional correspondence.[2]

Today the "Callendar Effect" refers to his theory that global climate change can be attributed to an enhanced greenhouse effect due to elevated levels of carbon dioxide in the atmosphere from anthropogenic sources, primarily from the combustion of fossil fuels. This is a complex story of discovery, well documented in Callendar's publications but hitherto largely unexplored in his personal papers and notebooks. His contributions were well known to his contemporaries, but have remained largely unnoticed and underappreciated by most scientists and the general public, and, surprisingly, also by historians of science and technology. Here is Callendar's story of the discovery of global warming and the role of anthropogenic CO_2.

The Nineteenth-Century Background

> Tyndall was the first to put forward the CO_2 theory of Ice Ages
>
> —Callendar[3]

The theory of the greenhouse effect has a long and distinguished intellectual ancestry that includes work done by Joseph Fourier, John Tyndall, and Svante Arrhenius, in the early, middle, and late decades of the nineteenth century.[4] There were many others. Arrhenius even introduced, briefly, the notion that humans might be global geophysical agents of climate change.

Jean Baptiste Joseph Fourier (1768–1830) is remembered today as a mathematical physicist, but he was known to his contemporaries as, variously, a secret policeman, a political prisoner (twice), governor of Egypt, prefect of Isère and Rhône, friend of Napoleon, baron, outcast, and perpetual member and secretary of the French Academy of Sciences, whose fortunes rose and fell with the swirling political tides.[5] Fourier considered himself the Newton of heat: "The principle of heat penetrates, like gravity, all objects and all of space, and it is subject to simple and constant laws."[6] For Fourier, solar heating and the temperature of outer space were the most significant factors controlling the Earth's heat budget.[7] He did, however write suggestive passages about what others later called the greenhouse effect. For example: "the temperature [of the Earth] can be augmented by the interposition of the atmosphere, because heat in the state of light finds less resistance in penetrating the air, than in repassing into the air when converted into non-luminous heat."[8] Yet the author of the *Analytical Theory of Heat* admitted that it is "difficult to know how far the atmosphere influences the mean temperature of the globe; and in this examination we are no longer guided by a regular mathematical theory."[9]

In the middle decades of the nineteenth century John Tyndall (1820–1893), an Irish-born scientist and consummate experimentalist at the Royal Institution of Great Britain, worked on the radiative properties of various gases, demonstrating that "perfectly colourless and invisible gases and vapours" were able to absorb and emit radiant heat. He identified the importance of atmospheric trace constituents as efficient absorbers of longwave radiation and as important factors in climatic control. Specifically, he established beyond a doubt that the radiative properties of water vapor and carbon dioxide were significant factors in explaining meteorological phenomena such as the formation of dew, the energy of the solar spectrum, and, possibly, the variation of climates over geological time.

An accomplished lecturer and writer, Tyndall employed numerous metaphors to describe his experiments with radiant heat. The following is reminiscent of the greenhouse effect:

> The solar heat possesses . . . the power of crossing an atmosphere; but, when the heat is absorbed by the planet, it is so changed in quality that the rays emanating from the planet cannot get with the same freedom back into space. Thus the atmosphere admits of the entrance of the solar heat, but checks its exit; and the result is a tendency to accumulate heat at the surface of the planet.[10]

Elsewhere Tyndall wrote: "The aqueous vapour constitutes a local dam, by which the temperature at the earth's surface is deepened; the dam, however, finally overflows, and we give to space all that we receive from the sun."[11]

The carbon dioxide content of human lungs and in regional air samples was also a focus of Tyndall's investigations.[12] On a more cosmic level, Tyndall thought that changes in the amount of any of the radiatively active constituents of the atmosphere—water vapor, carbon dioxide, ozone, or hydrocarbons—could have produced *"all the mutations of climate which the researches of geologists reveal* . . . they constitute true causes, the *extent* alone of the operation remaining doubtful."* Tyndall gave credit to his predecessors, including Fourier, for the intuition that "the rays from the sun and fixed stars could reach the earth through the atmosphere more easily than the rays emanating from the earth could get back into space." The experimental verification of this phenomenon, however, belonged to Tyndall.[13] His laboratory experiments with microphysical entities on London's Albemarle Street had informed a new view of the atmosphere.

In 1896 the Swedish scientist Svante Arrhenius (1859–1927), following Tyndall's suggestion, demonstrated that variations of atmospheric CO_2 concentration could have a very great effect on the overall heat budget and surface temperature of the planet and might trigger feedback phenomena that could account for glacial advances and retreats.[14] Arrhenius's model relied heavily on the experimental and observational work of others, including Josef Stefan's new law that radiant emission was proportional to the fourth power of temperature; Samuel P. Langley's measurements of the transmission of heat radiation through the atmosphere; Léon Teisserenc de Bort's estimates of cloudiness for different latitudes; Knut Ångström's early values for the bulk absorption coefficients of water vapor and carbon dioxide; Alexander Buchan's charts of mean monthly temperatures over the globe; and Arvid Högbom's ideas about the carbon cycle, in which volcanic eruptions were the chief source of carbon dioxide in the Earth's atmosphere. Note that Arrhenius did not write his 1896 essay because of any great concern for increasing levels of CO_2 caused by the burning of fossil fuels; instead, he was attempting to explain temperature changes at high latitudes that could account for the onset of ice ages and interglacials.

The notion that humans can influence climate has deep historical roots.[15] This idea appeared briefly in a new form in 1899 when Nils Ekholm, an associate of Arrhenius, pointed out that at present rates, the burning of pit coal could double the concentration of atmospheric CO_2 and could "undoubtedly cause a very obvious rise of the mean temperature of the Earth."[16] In his book *Worlds in the Making*, Arrhenius popularized Ekholm's observation, noting that "the slight percentage of carbonic acid in the atmosphere may by the advances of industry be changed to a noticeable degree in the course of a few centuries." Arrhenius considered it likely that in future geological ages the Earth would be "visited by a new ice period that will drive us from our temperate countries into the hotter climates of Africa." On the timescale of hundreds to thousands of years, however, Arrhenius speculated on a "virtuous circle" in which the burning of fossil fuels could help prevent a rapid return to the conditions of an ice age, and could perhaps inaugurate a new carboniferous age of enormous plant growth. Ekholm concurred, adding his own speculations about the possibility of deliberate climate modification to delay the onset of an ice age, for example by burning coal exposed in shallow seams.[17]

Early Twentieth-Century Doubts about CO₂

No possible increase in atmospheric carbon dioxide could materially affect either the amount of insolation reaching the surface or the amount of terrestrial radiation lost to space.

—*Climate and Man* (1941)[18]

Soon, however, the efficacy of CO_2 as an infrared absorber and factor in climate change was being challenged. The geologist T. C. Chamberlin, author of several papers on ice ages and interglacials, felt that the role of atmospheric CO_2 had been overemphasized and that not enough attention had been given to the oceans.[19] Knut Ångström concluded that CO_2 and water vapor absorb infrared radiation in the same spectral regions and that any additional CO_2 would have little or no effect on the climate.[20] Based on their research on the solar spectrum, Charles Greely Abbot and F. E. Fowle Jr. insisted on the primacy of water vapor as an infrared absorber.[21] In 1929, G. C. Simpson pointed out that it was "now generally accepted that variations in carbon-dioxide in the atmosphere, even if they do occur, can have no appreciable effect on the climate."[22]

A plethora of competing and speculative theories had been proposed over the years to account for secular changes of climate: changes in solar output; changes in the earth's orbital geometry; changes in terrestrial geography, including the form and height of continents and the circulation of the atmosphere and oceans; and changes in atmospheric transparency and composition, in part due to human activities. There were many others.[23] William Jackson Humphreys, meteorological physicist with the U.S. Weather Bureau and a strong proponent of the theory that volcanic dust was the leading cause of ice ages, thought that none of the current theories were adequate: "Change after change of climate in an almost endless succession, and even additional ice ages, presumably are still to be experienced, though . . . when they shall begin, how intense they may be, or how long they shall last no one can form the slightest idea."[24]

Callendar had read the early works on atmospheric absorption by Arrhenius, Ekholm, Ångström, and others but was not convinced. He commented that Ekholm's paper, "although a considerable advance on previous work of the same kind," had "made use of the rather unreliable values then

available, of CO_2 and H_2O absorption, to determine the atmospheric radiation in various air thicknesses." Callendar was referring to the use of values
for the absorption by CO_2 of the whole beam radiation from a blackbody at
various temperatures. He observed that "the values and methods used render
these [measurements] out of date."[25]

Callendar's notebooks are filled with comments of both technical and
historical interest on the infrared properties of atmospheric gases (Figure
5.1). One early entry concerns "the first spectrum measurements in the infrared," of CO_2 and H_2O bands from 2 to 5 μ published in 1892 by the German spectroscopist L. Paschen.[26] Another concerns the attempt by E. O.
Hulbert, a physicist at the U.S. Naval Research Laboratory, to account for
the atmospheric temperature distribution using approximate absorption
coefficients given by earlier authors. Although Hulbert did not address the
issue of anthropogenic carbon dioxide, Callendar noted that this 1931 paper
supports the CO_2 theory of ice ages and shows that "it is at least a possible
one."[27] There are many more such entries in Callendar's notebooks.

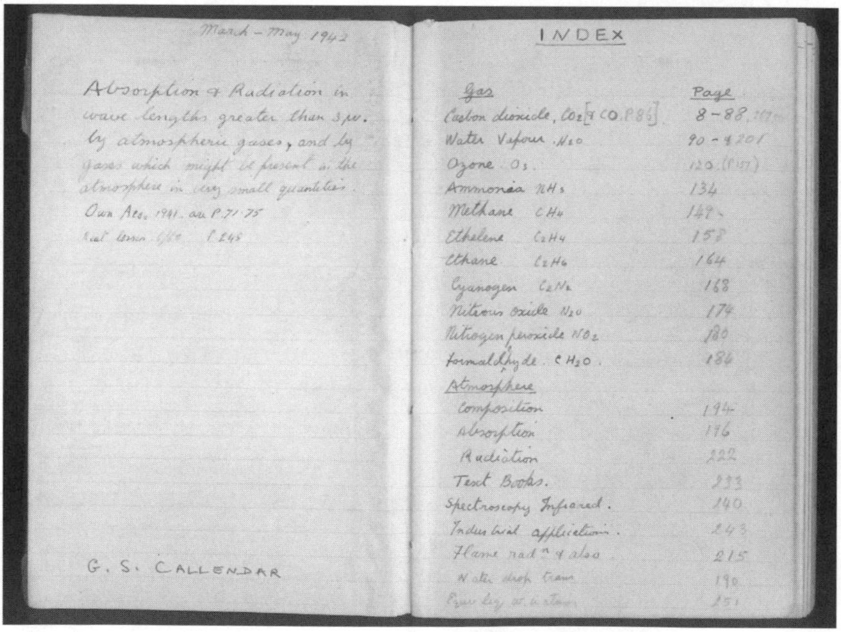

Fig. 5.1. First two pages of Callendar's notebook on absorption and radiation by atmospheric gases, one aspect of his defense work during World War II (see Chapter 4). CP 2,
Notebook 1942_IRS.

Callendar's Climate Publications, 1938–1964

In the prime of his research years, Callendar published a remarkable series of essays on atmospheric warming and anthropogenic CO_2.

1938

In his first and most widely cited article on climate change, "The Artificial Production of Carbon Dioxide and Its Influence on Temperature" (1938), Callendar provided scientific links between rising fuel combustion, rising levels of atmospheric CO_2, infrared heating by trace gases, and the rise in world temperatures in recent decades. He estimated that humans have added some 150,000 million tons of carbon dioxide to the air since the end of the nineteenth century through the combustion of fossil fuels, with approximately three-quarters of this remaining in the atmosphere.[28] Using the latest information on the radiation absorption coefficients of carbon dioxide and water vapor, he calculated that enhanced "sky radiation" from the artificial production of carbon dioxide was one of the major factors causing the rise of mean global temperatures as measured by 200 weather stations worldwide.

In the paper, Callendar's first goal was to establish that fuel burning had exceeded the limits of the natural carbon cycle and that the oceans, which act as a "giant regulator of carbon dioxide," would not be able to absorb all or even most of the excess. His preliminary results (later updated) indicated that the concentration of CO_2 in the atmosphere had risen from a value of 274–292 parts per million (ppm) in 1900 (which he considered the nineteenth century background) to a current value of 289–310 ppm and was expected to increase about 10 percent or more per century thereafter.

The core of the paper is his presentation of techniques for calculating the radiative effect of carbon dioxide in three primary bands: 2.4–3 μ, 4–4.6 μ, and 13–16 μ. To this Callendar added information not available to earlier researchers on the vertical thermal structure of the atmosphere, the pressure broadening of absorption lines, and the detailed infrared spectrum of water vapor. He calculated the vertical component of sky radiation by dividing the air into horizontal layers of known mean temperature and trace gas content and summing the absorbing powers of each layer in the different wave bands—this for various surface temperatures and assumed atmospheric conditions from the poles to the tropics. An increase in the carbon dioxide content of the atmosphere lowers the altitude and thus raises the effective radiating temperature of the sky radiation reaching the ground. The result is net radiative heating of the Earth's surface. Callendar estimated that two-thirds of the warming trend of 0.005°C per year during the past half-century was due to the artificial production of carbon dioxide. His model predicted a 2°C global warming for doubled carbon dioxide concentrations.[29]

Callendar was by no means an environmental alarmist, since he wrote of the benefits of the combustion of fossil fuels, including extension of the cultivated region northward, stimulation of plant growth by carbon dioxide, and the probable indefinite delay of the "return of the deadly glaciers." He ended his paper with a prediction that "the course of world temperatures during the next twenty years should afford valuable evidence as to the accuracy of the calculated effect of atmospheric carbon dioxide." This prediction, published in 1938, would be more widely cited today if he had written "during the next fifty to one hundred years." This is because a relatively brief cooling trend began in the 1940s and extended to the end of his life. A large exclamation point (!) marks this passage on page 236 of Callendar's own copy of the paper.[30]

Convincing others about global warming was an uphill battle. Few of his scientific colleagues could imagine that human activities had any influence on "phenomena of so vast a scale" as climate. Yet he hoped to show that "such influence is not only possible, but is actually occurring at the present time." During the discussion of this article at the Royal Meteorological Society, G. C. Simpson, who advocated a theory based on changes in solar radiation, pointed out that the atmosphere was not in a state of radiative equilibrium and that convection and other air movements would have to be taken into account. Simpson regarded the recent rise of CO_2 content and temperature as coincidental and pointed to other complicating factors. Callendar replied that the atmospheric heat budget was extremely complex but noted, "If any substance is added to the atmosphere which delays the transfer of low temperature radiation, without interfering with the arrival or distribution of the heat supply, some rise of temperature appears to be inevitable in those parts which are furthest from outer space." In other words, *the greenhouse effect is real.*

Callendar responded to F. J. W. Whipple's question about the "natural movements of carbon dioxide" (i.e., the carbon cycle) by saying that he had actually written such an account, but had omitted it because of space limitations. David Brunt thought that the recent change in the mean temperature was no more striking than the changes that appear to have occurred in the latter half of the eighteenth century, and L. H. G. Dines thought it could merely be a "casual variation." Brooks disagreed with both of his colleagues, saying that "he had no doubt that there had been a real climatic change during the past thirty or forty years," but he was not willing to attribute this to anthropogenic carbon dioxide. Callendar replied that past historical warm periods appeared to be short warm intervals of up to 10 years, with some very cold years intervening, whereas the widespread recent warming was more "gradual and sustained" over a 40-year period. He conceded to Brooks that the recent rise in arctic temperatures was far too large to be attributed solely to changes in CO$_2$ concentration, but he thought that such a mechanism might "act as a promoter to start a series of imminent changes in the northern ice conditions."

J. H. Coste called into question the reality of increases in atmospheric CO$_2$ and thought it was "very doubtful whether the differences which Mr. Callendar made use of were real." Callendar defended the accuracy of the early CO$_2$ measurements, claiming he had only used values observed on days when strong and steady west winds were blowing at Kew. The actual CO$_2$ added to the atmosphere by fuel combustion in the last 40 years was equal to an increase of 8 percent, while "the observed and calculated values agreed" in giving an effective increase of about 6 percent.[31]

Although this was Callendar's first paper presented to the Royal Meteorological Society, it is clear from the discussion that he was accustomed to rubbing elbows with scientific elites and was already a seasoned veteran of scientific debate. In this his work with his father and Egerton on steam tables had served him well. Callendar published three other shorter papers in 1938, all on the subject of rising temperature trends in the British Isles and worldwide.[32]

1939

Today most people are familiar with announcements, usually published in January, that the previous year (or decade) was the warmest on record. Callendar published a similar note in 1939 reminding the Royal Meteorological Society that the previous year had been anomalously warm, and that "in most North Atlantic countries the mean temperature of last year was remarkably high and at several stations in North America, Scotland, and the Baltic countries it equaled or exceeded previous record values."[33]

Callendar strengthened this claim in his second climate-related article, "The Composition of the Atmosphere through the Ages," written at the invitation of the Air Ministry.[34] Here he wrote, "the five years 1934–38 are easily the warmest such period at several stations whose records commenced up to 180 years ago." This article contains an early statement of the now-familiar claim that humanity has become an agent of global change:

> It is a commonplace that man is able to speed up the processes of Nature, and he has now plunged heavily into her slow-moving carbon dioxide into the air each minute. This great stream of gas results from the combustion of fossil carbon (coal, oil, peat, etc.), and it appears to be much greater than the natural rate of fixation. . . .[35]

Exceeding the accuracy of the available measurements, but intuitively feeling he was right, Callendar reported a 7 percent increase in anthropogenic CO_2, "from 0.028 per cent about the year 1900 to 0.030 percent of recent years" and linked the 1°F rise in temperature from 1900 to 1938 to the concurrent increase in industrial emissions of carbon dioxide. In effect, humanity was conducting a grand geochemical experiment:

> As man is now changing the composition of the atmosphere at a rate which must be very exceptional on the geological time scale, it is natural to seek for the probable effects of such a change. From the best laboratory observations it appears that the principal result of increasing atmospheric carbon dioxide . . . would be a gradual increase in the mean temperature of the colder regions of the earth.[36]

1940–1944

We know that Callendar was very busy in the early years of World War II finishing his steam research, seeking a permanent job, and relocating his family. Still, he found time to write a review of the history and present state of knowledge of observations of atmospheric carbon dioxide, in support of his claim that "the amount of this gas in the air has increased of late years." He also addressed the immense but slow-acting capacity of seawater to absorb CO_2, but thought it will "doubtless take many centuries to stablise the great eruption of this gas, now about 300 million cubic metres per hour, which has resulted from human activities."[37]

Callendar's most significant article of 1941 reviewed spectroscopic measurements of the absorption bands of CO_2, the effect of pressure broadening on line widths, and the meteorological importance of atmospheric radiation.[38] Recall that the carbon dioxide theory of Tyndall and Arrhenius that was based on graybody absorption had been moribund for four decades. Now, however, atmospheric radiation was receiving renewed attention, mainly for its role in the formation of cold air masses. Although water vapor had received most of the attention, the radiative properties of carbon dioxide were also being reconsidered.[39] Callendar gathered together the observational data on CO_2 that was scattered through the scientific literature and had been published in several languages over several decades. He also interpreted experimental data collected for different wave bands at various temperatures and pressures, reducing it to a form usable to meteorologists. The result, clearly illustrated in the paper for an atmospheric temperature of 263 K, showed CO_2 absorption bands partly overlapping, but distinct from those of H_2O, N_2O, and O_3. It also depicted atmospheric infrared "windows" centered at 10 and 18 μ. All this fit well with Callendar's stated research agenda, which was to "reconsider the difficult problem of the effect of changes in the amount of carbon dioxide on the temperature of the atmosphere with the aid of the much more accurate absorption values given here."[40]

Callendar's paper drew rave reviews and changed the opinions of such notable scientists as David Brunt and Sydney Chapman. Brunt was convinced that CO_2 absorption was "rather more important that had been thought in the past" and would play a major role in future study of the heat balance of the atmosphere. Chapman admonished his fellow meteorologists for not collecting their own data sooner on this important subject and for relying on investigations by spectroscopists and other physicists. He suggested that meteorologists conduct an organized research program on atmospheric radiation. In reply, Callendar thanked the "experienced meteorologists" for their kind assessments and expressed hope that strong financial support over a multiyear period might be forthcoming for the needed measurements. Remember, Callendar was looking for a job in 1941.

Thinking globally during the desperate war years of 1941 and 1942 was not a luxury for most individuals, yet Callendar managed to publish short notes on the ocean's influence on weather, on the movement of glaciers in response to changes in air temperature, and on a quest for a "reliable climatic indicator" for the Southern Hemisphere of comparable accuracy to the numerous temperature and glacial observations of the north.[41] Undoubtedly, Callendar's most important paper of the war years was his study, published jointly with Sutherland, on the infrared spectra of atmospheric gases other than water vapor (see Chapter 4).[42] The authors sought to advance detailed quantitative knowledge of the vibration-rotation spectra of these molecules, including phenomena induced by elevated temperatures such as Doppler broadening, collisional broadening, and the populating of higher energy levels at increased temperatures. In a comment clearly targeted at members of the Gassiot Committee of the Royal Society chaired by G. M. B. Dobson, the authors called for a "vast amount" of additional work, largely experimental, to be conducted on the following issues:

a. Accurate measurements on the absolute absorption coefficients in the principle bands of CO_2, O_3, and N_2O at a series of different pressures and temperature and in the presence of varying amounts of water vapor.
b. Accurate measurements of the absolute absorption coefficients for at least one band of each of the following: NH_3, CH_4, C_2H_4, C_2H_6, C_3H_8, C_2N_2, HCN, N_2O_4, N_2O_5.
c. Accurate plot of the atmospheric spectrum between 2 and 7 μ, using a slit width of 1 cm^{-1} or less on a grating spectrometer.
d. Estimates of the relative importance of various absorption bands of the known and the fairly probable atmospheric gases, in the absorption and emission of radiation.

Clearly, the study of radiation in the atmosphere was taking a great leap forward.

Although involved full time in classified defense work at Langhurst, Callendar continued to publish on climate change, even when printing and paper shortages forced *Meteorological Magazine* to issue their 1943 volume in typescript. His brief contributions in 1943 and 1944 discussed climate change in the north and the variations of winter temperature over the past eight centuries.[43]

1947–1955

Following the war, Callendar, who was always reevaluating and revising his own work, published a brief note in the *Quarterly Journal* to bring new work on atmospheric radiation to the attention of meteorologists.[44] Specifically, he challenged the widely held assumption, promoted by Elsasser's radiation diagram, that the amount of CO_2 in one thickness of the atmosphere can absorb all the energy of which it is capable. He also promised a revision of the CO_2 absorption calculations used in his model of 1938.

In his popular article, "Can Carbon Dioxide Influence Climate?" Callendar reported a steady rise in observed CO_2 content from its nineteenth-century background amount of 290 ppm and noted that the rate of CO_2 increase had been accelerating recently, due to the expansion of industry and the removal of forests. Given the 10 percent rise in CO_2 recorded from 1900 to 1935, Callendar would not have been surprised to learn that its atmospheric concentration had risen about 30 percent by the end of the twentieth century. Callendar wrote for the public:

> Reduced to its simplest terms this theory depends on the fact that, whereas carbon dioxide is almost completely transparent to solar radiation, it is partially opaque to the heat which is radiated back to space from the earth. In this way it acts as a heat trap, allowing the temperature near the earth's surface to rise above the level it would attain if there were no carbon dioxide in the air. . . . It may be said that the climates of the world are behaving in a manner which suggests that slightly more solar heat is being retained in the atmosphere. This could be due to its increasing opacity to terrestrial heat as a result of the additions of carbon dioxide.[45]

That was written in 1949!

In the early 1950s Callendar published fundamental research on the movement and dimensions of glaciers in response to temperature variations. He also investigated local temperature controls and fluctuations in the long-term records in England and Canada.[46] During this period, he corresponded with the infrared physicist Gilbert Plass (see below) and commented on Plass's paper in the *Quarterly Journal*.[47] Callendar noted with satisfaction that most of the points made in his 1949 paper were reaffirmed by E. W. Hewson and Plass at the Toronto Meteorological Conference in 1953.[48]

Global Warming Reaches the Public

Global warming was on the public agenda in the late 1940s and early 1950s, as Northern Hemisphere temperatures reached an early-twentieth-century peak. Concerns were being expressed by scientists and in the popular press about changing climates, rising sea levels, loss of habitat, and shifting agricultural zones. Hans Ahlmann, a noted geographer reported in 1948 that Iceland had recently experienced a 1.3°C warming and its glaciers were in retreat.[49] In 1950, MIT meteorologist Hurd C. Willet told the Royal Meteorological Society that the global temperature trend was "significantly upward" since 1885, with most of the warming occurring north of the 50th parallel.[50] Subsequent studies confirmed that from 1890 to 1940, the mean thickness of Arctic ice decreased by about 30 percent, and the areal extent of ice decreased by as much as 15 percent. At the same time the intensity of the global circulation increased markedly and the Earth became warmer—reportedly 10° warmer in the Norwegian Sea.[51] Evidence that the U.S. climate was warming was presented at the May 1950 meeting of the American Meteorological Society by Richmond T. Zoch of the Weather Bureau, who reported that Washington, D.C. had warmed about 3.5°F since 1862 when temperature records began. John H. Conover of Harvard reported a similar 3.0°F rise in the 100-year record kept at the Blue Hill Meteorological Observatory in Massachusetts (see Figure 6.1).[52]

The popular press followed the lead of the scientists. In 1950, the *Saturday Evening Post* asked, "Is the World Getting Warmer?" The article cited rising sea levels, shifts of agriculture, the retreat of glaciers, displacement of ocean fisheries, and the possible migration of millions of people displaced by climate warming. Of primary concern was the unprecedented rate of change.[53] *Time Magazine* reported in May 1953 that industrial CO_2 emissions could increase the atmospheric concentration by about 50 percent, causing a 1.5°F warming in 100 years, or possibly more if positive cloud feedback effects occur.[54] The article cited the authority of Gilbert Plass. Two months later, the *New York Times* Sunday magazine published a synopsis of the recent climate warming trend and a review of various theories—including the enhanced greenhouse effect—citing Callendar's earlier figure of a 10 percent rise in CO_2 level likely causing further warming and possible future problems with rising sea levels.[55]

As Callendar had pointed out in 1938, fuel burning had probably exceeded the limits of the natural carbon cycle, and the oceans would not be able to absorb all or even most of the excess. He warned of this again in 1939 when he wrote that the great stream of gas resulting from the combustion of fossil carbon "appears to be much greater than the natural rate of fixation." Thus, it was not really news in 1956 when *Time Magazine* reported a growing trend in annual CO_2 emissions, with possible dramatic consequences if the greenhouse effect intensifies. This was merely an update and acknowledgment of the Callendar Effect.[56]

Problematizing the CO_2 Theory

G. S. Callendar would be a better-known figure today if the climate had continued to warm monotonically (there was a noticeable cooling trend in the 1960s and 1970s), if he had headed a major research institution, and if scholars (especially historians) had looked beyond the temporal horizon of the International Geophysical Year (IGY) of 1957/58.[57] As it stands, Callendar influenced the American agenda regarding climate warming and CO_2 and helped shape the research program both in his day and in the decades following.

Callendar welcomed the prospect of new measurements to be taken during the IGY and offered investigators information and encouragement. Callendar was especially pleased that Charles D. Keeling's new technique would be employed and noted that the CO_2 concentrations of 315 ppm reported by Keeling from his mountain, desert, and coastal series were in "quite good agreement" with the 321 ppm average for similar locations in Norway in 1955 and 1956. He thought the average from the IGY Antarctic station should be "specially interesting" and might provide a "general free air value for the southern zone."[58]

Also on the horizon at the time of the IGY was a new technique for measuring radiocarbon abundances developed by Willard Libby and applied by Hans Suess. An apparently large discrepancy between the decrease of ^{14}C in the biosphere from the uptake of fossil fuel carbon (the "Suess Effect") and the increase in atmospheric CO_2 (the "Callendar Effect") led to a scramble to explain the carbon cycle.[59] A series of interrelated papers in the January 1957 issue of *Tellus* addressed this problem.[60] Roger Revelle and Hans Suess thought that the Callendar Effect was "quite improbable" on its own and was likely augmented by a number of factors, including a slight increase of ocean temperature, a decrease in the carbon content of the biosphere (a factor discussed by Callendar), and possible chemical and organic changes in the oceans. They also called into question Plass's calculation of a 0.36°C warming from CO_2. At the time uncertainties in carbon reservoirs, fluxes, and residence times were huge, ranging over seven orders of magnitude. The question of the timescale of ocean-atmosphere exchange, the deep ocean circulation, while widely discussed, was far from settled. The figures used by Revelle and Seuss were 100 times larger than those reported by Plass in 1956, yet 10,000 times smaller than those deduced by H. N. Dingle in 1954.[61] This fact severely limited their conclusions.

Rather than warning against the dangers of projected large increases in fossil fuel burning, Revelle and Suess recommended more detailed scientific study of the carbon cycle and how carbon dioxide is partitioned between the atmosphere, the oceans, the biosphere and the lithosphere, pointing to current uncertainties and new work that needed to be done:

> Present data on the total amount of CO_2 in the atmosphere, on the rates and mechanisms of CO_2 exchange between the sea and the air, and between the air and the soils, and on possible fluctuations in marine organic carbon are insufficient to give an accurate base line for measurement of future changes in atmospheric CO_2. An opportunity exists during the International Geophysical Year to obtain much of the necessary information.

Mark Bowen calls the article "basically a grant proposal" for the IGY and writes that "Revelle and Suess's paper used a tiny bit of new experimental information to come to a . . . previously demonstrated conclusion"—that is, ultimately, "Revelle and Suess agreed with Callendar and Plass." It is short-sighted, however, to ignore Revelle's earlier interest in carbon dioxide and radioisotopes, especially ^{14}C, dating back to early nuclear tests. Revelle also supported the work of such pioneers as Willard Libby, Harmon Craig, and especially Hans Suess, who established the La Jolla Radiocarbon Laboratory at the Scripps Institution of Oceanography.[62]

Callendar's response to this article, as recorded in his notebooks, letters, and publications, was measured and analytical. He was puzzled by the contradictions in the article as he struggled to digest it, realizing that it would probably lead to new measurements he desired. He wrote in his notebook, "Revelle and Suess say 'it seems quite improbable that the CO_2 increase could be ten percent this century as Callendar's statistical analysis indicates.' Then they list 'possible other factors which could have caused such a large figure. . . .'" He wrote in a letter to Plass, "A point of special interest is the large discrepancies between the apparent increase of atmospheric CO_2 given by the air-CO_2 observations ... and the predicted increase derived from the size of the exchange reservoirs as now revealed by radio carbon measurements." He wrote in *Tellus* that the biosphere and warm weather conditions could have skewed some of the early CO_2 measurements toward the high side, but he stood by his nineteenth century background value of 290 ppm and a rising trend in atmospheric CO_2 "which is similar in amount to the addition from fuel combustion." Callendar admitted that his result was not in accordance with recent radio carbon data, but pointed out that the reasons for the discrepancy were obscure and that much more observational data would be required to clarify this problem. He was especially eager to see that new measurements were to be taken at high latitudes in the Southern Hemisphere during the IGY.[63]

Plass wrote to Callendar that Revelle and Suess and Arnold and Anderson had "*attacked the carbon dioxide climatic theory quite vigorously* at a meeting earlier this year."

> They claimed that it was absolutely impossible to have had a sufficient increase in the CO_2 amount in this century for the reasons that were given in their articles. I think you have pointed out several ways that their conclusion could be in error and I feel that there are still several possible explanations.[64]

The resolution came in 1958 in a publication by two Swedish scientists, Bert Bolin and Erik Eriksson, titled, "Changes in the Carbon Dioxide Content of the Atmosphere and Sea due to Fossil Fuel Combustion", which clarified the buffering effect of sea water when changes of the partial pressure of CO_2 take place.[65] Their calculations supported Callendar's estimate of a 10 percent increase by 1956 in atmospheric CO_2 and predicted a likely increase in this value of about 25 percent by the year 2000—a very accurate prediction even before the results of Keeling's measurements were known.

The Final Publications
Callendar spent much of his time studying temperature fluctuations and trends around the world, examining the location and situation of some 400 stations, and discussing the reliability of the observations. He charted

regional and zonal fluctuations, and he took into account the homogeneity of the records, including evidence of glacial recession in different zones and the significance of station relocation and the growth of cities. Of the 96 research notebooks in the collection at the University of East Anglia and on the Callendar Papers DVD (available from the American Meteorological Society), about 60 percent are filled with world and regional temperature data.[66]

In 1961 he published the results of his study in the *Quarterly Journal*, concluding that the pattern of recent climatic warming was not incompatible with his hypothesis of increased carbon dioxide radiation."[67] In published comments on the article, prominent American climatologists Helmut Landsberg and J. Murray Mitchell noted rather close agreement between Callendar's results and a similar study by Mitchell and Hurd Willet, but challenged Callendar on his dismissal of other causal mechanisms and his assumption that CO_2 released in the Northern Hemisphere was slow to mix globally.

As this paper was going to press, Callendar wrote a note listing "[Four] reasons for the unpopularity of CO_2 theory in some meteorological quarters." Although there was no organized opposition to anthropogenic climate change at the time, Callendar's note reads much like a contemporary response to global warming skeptics:

a. The idea of a single (easily explained) factor causing world wide climatic change seems impossible to those familiar with the complexity of the forces on which any and every climate depends.
b. The idea that man's actions could influence so vast a complex [system] is very repugnant to some.
c. The meteorological authorities of the past have pronounced against it, mainly on the basis of faulty observations of water vapour absorption, but also because they had not studied the problem to anything like the extent required to pronounce on it.
d. Last but not the least. They did not think of it themselves![68]

In March 1964, Callendar wrote himself a note on "Reasons why this paper has proved unwelcome to some" listing the following factors:

a. The introductory sections . . . are equivalent to a lecture on how this sort of climatic data should be presented for the general scientific reader. By contrast most papers on this subject appear clumsy and confusing—It is most annoying [to others] to have the shortcomings of one's own work made so evident.

b. The presentation is ahead of previous work on temperature series and this makes the latter workers feel relatively ignorant, which is irritating.

c. The climatic/circulation factors are dismissed in half a dozen lines—as these . . . factors have been the subject of a vast amount of speculation in innumerable papers, this summary treatment is very annoying to the writers.

d. CO_2 as a cause of climatic change . . . is above the heads of nearly all writers on the subject, to whom it is just another speculation on a par with solar—u.v.—ozone effects.[69]

Callendar understood that he had been presumptuous, adopting a "school teacher" attitude," "laying down the law on how it should be done," and reacting to the "exceptionally low and muddled quality" of most papers on climatic subjects. Such didactic excesses may find a place in some classrooms, but not in a scholarly journal, and Callendar paid a price for this. Yet he felt "fully justified" by his lifelong efforts and working mastery of instrumental data, temperature trends, infrared physics, and the behavior of CO_2.

Callendar's final publication was a short exchange with a bird watcher, G. Harris, on how the recent climate cooling trend was affecting European birds. While Callendar warned of possible computational errors in the temperature series and changes in station location as reasons for the assumed cooling trend, Harris was sure that Callendar would agree with the picture of "a general decline of temperature in recent years."[70] Unquestionably, the global climate was cooling in the 1960s. The bitter winter of 1962–63 was the coldest on record in England and Wales since 1740. The noted climatologist Hubert H. Lamb, one of Callendar's correspondents, has written about the "clustering" of harsh winters, referring to a sequence of three "skating Christmasses" in England with very severe frosts or snow in 1961, 1962, and 1963. Such frosty or white Christmases in England were exceedingly rare in the preceding half century.[71] Callendar must have been pondering his scientific legacy on climate warming as he was shoveling the walkways during the record snowfall of December 1962 (Figure 2.12) and shivering through his final December in 1963.

Climate and Carbon Dioxide

From his retirement in 1958 to his death in 1964 Callendar was working diligently on a book on the carbon cycle with the working title "Climate and Carbon Dioxide." He never completed this project. All that remains is a

FRONTISPIECE
Putting Carbon on the air.
A power station such as this can add as much as 300 million cub feet of Carbon Dioxide to the air each day.

Fig. 5.2. Frontispiece for "Climate and Carbon Dioxide" (the book was never published).

tantalizing list of figures and plates to be inserted into a manuscript of at least 18 chapters and over 176 pages (see Figures 5.2, 5.3, and 5.4). This list indicates chapters on anthropogenic sources, the biosphere, lithosphere, volcanoes, oceans, ice ages, and the increase in temperature due to CO_2 production.

List of Plates for "Climate & Carbon Dioxide.

Plate No.	Title.	Opp page of MS no.	chapter.
1.	Ancient coal forming swamp & forest.	6 or 7.	1.
2.	Anticline in dark carbonacious shale.	23.	3.
3.	Vesuvius in eruption.	32.	5.
4.	Calcium Carbonate precipitated from sea water.	76.	10.
5.	Foriegn boulder transported by ancient glacier.	176	18

Frontispiece. Putting Carbon on the air.

List of Figures for "Climate & Carbon Dioxide.

Fig No.	Title	opp page of MS no.	chapter of MS
1.	Increase of atmospheric CO_2 pressure.	84.	11.
2.	Quantities of CO_2 & Water Vapour.	89,	12.
3.	The effect of atmospheric CO_2 on surface temperature.	172	18.
4.	Increase of temperature due to CO_2 production.	174.	18.

Fig. 5.3. List of plates (top) and figures (bottom) for "Climate and Carbon Dioxide" indicating a manuscript of over 176 pages with 18 chapters.

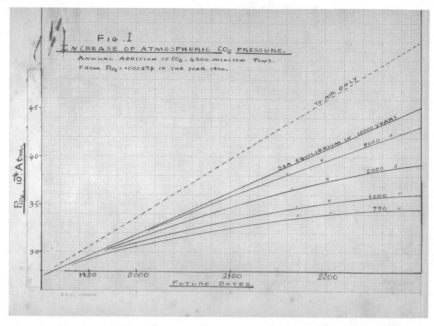

Fig. 5.4. Callendar's Figure 1, Increase of atmospheric CO_2 pressure for various assumptions about ocean exchange.

It is possible that a complete manuscript or additional fragments will be uncovered someday, but this is not likely. Anne and Bridget's inventory of their father's papers (Appendix C) and Anne's letters to Derek Schove just after Callendar's death in October 1964 do not indicate its existence.

Conclusion

> CO_2 has begun to come home, hasn't it!
>
> —Manley (1958)[72]

The carbon dioxide theory of climate change was in deep eclipse when Guy Stewart Callendar established it on a firm scientific basis in 1938. In the next 25 years he published 10 major articles and over 25 shorter pieces on global warming, infrared radiation, and anthropogenic CO_2, working diligently, without support or institutional resources, to communicate his results to scientists and the educated public. His intensive research and sound scientific judgment led him to conclude that the trend toward higher temperatures was significant, especially north of the 45th parallel; that increased use of fossil fuels (and, to a lesser extent, deforestation and other practices) had caused a rise of the concentration of CO_2 in the atmosphere that would only continue and would most likely accelerate; and that increased sky radiation from the extra CO_2 was directly linked to the rising temperature trend. The world's glaciers were shrinking and he thought it unlikely that the oceans would absorb the excess carbon.

He scratched his head at the time of the IGY when scientists like Revelle and Suess and Arnold and Anderson seemingly confused the issues, but rejoiced when they and other colleagues such as Plass and Keeling brought new resources, analytical techniques, and theoretical insights to bear on the problem. Callendar passed away during a cooling trend in the 1960s, but most scientists today would agree that his early statements have been verified. One author recently referred to Callendar's insights as "pretty much spot-on" and his work on the enhanced greenhouse effect and human agency as being "roughly half a century ahead of his time."[73] The legacy of his articles, letters, and notebooks now passes to a new generation of scientists and historians living in a slightly warmer world with levels of atmospheric carbon dioxide unknown to Callendar.

CHAPTER SIX

Callendar's Legacy

Callendar—on Oct. 3 at 44 Parsonage Road, Horsham. Guy Stewart aged 67. Cremation took place at Worth on Tuesday, Oct. 6.

—*West Sussex County Times*[1]

This brief obituary notice marked the sudden passing of Guy Stewart Callendar. Letters of condolence from Callendar's scientific correspondents arrived within a fortnight. The Glaciological Society conveyed its official sentiments.[2] A letter from Gilbert Plass to Phyllis was more personal and addressed his scientific legacy:

> I was very grieved to receive your recent letter advising me of the death of your husband. Although I had never met him personally, we had corresponded extensively through the years and I felt as though I had known him. His contributions to the carbon dioxide problem in the atmosphere and to many other meteorological questions will always be remembered.[3]

The noted climatologist Gordon Manley wrote a letter of condolence on behalf of the "loss that many meteorologists will feel":

> In what Sir Graham Sutton has described as "the most deficient science in the world" your husband made a most original and lively contribution. I heard it myself in 193[8]; and when he retired he was quite indefatigable in slogging out the figures on which his—and my—arguments depend. I always enjoyed his letters for their crisp content and I can assure you that in the Royal

Meteorological Society we have a very high esteem for the few scientists who can match up to, and challenge, the "professionals" of the Meteorological Office. . . . I should like to assure you how that classic paper by G. S. Callendar in "Q.J. Royal Met. S. 193[8]" stands as an original contribution nearly twenty years before the Americans began to think along similar lines. There is a permanent stock in science to which we may all hope to add; and your husband [gained our] esteem indeed for doing just that.[4]

One of Callendar's frequent correspondents, schoolmaster Derek Justin Schove of St. David's College, wrote immediately expressing his shock at the news and his sympathy for the family: "I have for a long time been a great admirer of his work and we have been working together on temperature series. If there is anything I can do to help sort out his papers on that subject do please let me know."[5]

The Papers

There were indeed practical arrangements to be made concerning Callendar's instruments, papers, books, and research documents. Anne took the lead in these tasks, working with Bridget to sort the materials and filling a notebook with a complete inventory of their father's study at the time of his death.[6] Their extensive lists (see Appendix C) included:

Books and magazines
Reprint copies of published papers
Unpublished papers
Notebooks on "world climate and temperature figures"
Instruments
Correspondence
Loose papers and figures (see, for example, Figure 6.1)

Of considerable interest are the following unpublished papers listed by Anne and Bridget, but not in the collection: "A Contribution to the Problem of Glacial Climates, The Carbon Dioxide Theory," 1953 and "The Exchange of Carbon Dioxide Between the Sea and the Atmosphere," 1954. The latter was apparently written some three years before the IGY. Fortunately, the charts prepared for this study *are* in the collection (see, for example, Figure 5.4).[7]

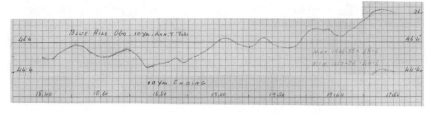

Fig. 6.1. Temperature trend at Blue Hill Observatory, 1840–1960, an example of one of Callendar's many hand-drawn figures.

Also in the collection is Callendar's manuscript, "Note on the Spectrum of Outgoing Radiation from the Troposphere." This paper demonstrates his understanding of the behavior of trace gasses at various environmental temperatures and pressures through the atmosphere.[8]

Former institutions and associates assisted the family. The physics department at Imperial College offered to purchase some of Callendar's personal instruments.[9] The Royal Society of London was "very pleased" to receive 100 reprints of Callendar and Egerton's 1960 study on steam, "as the demand for your father's paper was much greater than anticipated and it has been out of print for some time."[10] Anne offered Schove the full use of the materials:

I enclose a list of my Father's papers and books. I realize that some of it will not be of interest to you, particularly the reference books and magazines, but I thought it best to send you the complete list—then you can let us know what you would like from amongst the lot. We are mainly concerned that his notebooks should go to someone who will be able to make full use of their contents, because we know how much time and thought he spent in collecting all the figures and information in them. I'm sure you know how careful he was to produce accurate climatic data, and some of the notebooks, without doubt, contain valuable temperature series and weather records. Some are rather "rough" of course and may be earlier material and of less interest. I would like to have sent greater detail about the contents of the notebooks, but found it an impossible task to even try to list what was in each one, so have sent a very general summary—no doubt you will have a better idea yourself of the wealth of material in them. If by any chance you don't want them yourself, perhaps you would advise us of any society (or possibly another meteorologist) who would find them valuable. We shall be pleased to send you any of the things in the list that you would find useful and helpful in your research.[11]

For almost 20 years, the Callendar Papers were in the possession of Schove at St. David's College, Kent.[12] Schove, whose interests lay in climate data and sunspot cycles, kept in touch with Anne and even suggested she take up climatology like her father. He attempted some plots of Callendar's data and kept the papers in relatively good order. With Anne's permission, Schove lent at least one notebook on CO_2 to Charles D. Keeling at Scripps (since returned, see below) and dispersed extra copies of reprints to interested colleagues.

International interest in the environment was growing rapidly in the mid- to late-1970s. Lester Machta, member of the Scientific Committee on Problems of the Environment (SCOPE) and director of the Air Resources Laboratories at NOAA,[13] sent an inquiry on behalf of Keeling to the British Meteorological Office seeking the use of Callendar's papers, especially on the question of the pre-1900 CO_2 concentrations. The message found its way to Anne:

> As you probably know there is currently much interest in scientific circles on the possible impact of increasing atmospheric carbon dioxide on climate. Because of this, considerable attention is focused on the detail of your late father's pioneering work on the subject especially as regards his analyses of the early CO_2 records. . . . Your father's papers may reveal detail of the early CO_2 concentration that could help us decide the contribution that the burning of forests for agriculture was making to the atmospheric CO_2 concentration (in addition to the burning of fossil fuels) at the turn of the century and so help us in detailing the lifetime of CO_2 in the atmosphere. These are very important factors when we attempt to forecast CO_2 trends in the future. . . .
>
> Your father's published papers are all very well known. It is by studying these that there is a feeling that he may have found early CO_2 data which were never published. It was very prudent of you to put his papers in safe keeping. I do hope that they are still available.[14]

In 1984 Schove reported to Anne that he had had a "visit from an American—from La Jolla" and had lent him some important unpublished work on CO_2. The visitor was Charles D. Keeling, who borrowed the 1939/40 notebook on CO_2 and used it in a 1986 publication.[15]

About this time Schove suggested to Anne that the papers be put on "semi-permanent loan to a library or archive." Possible repositories included the following:

a. The Meteorological Office at Bracknell, which "now appears to be looking after its archives and has full library facilities and they are willing to accept the Callendar and my other meteorological books and papers."
b. The Climatic Research Unit (CRU) at Norwich under Tom Wigley, but they "have not so far expressed much interest in them. Their present library is small (but select)."
c. The Geography Schools at Oxford where Schove's daughter was a don.
d. The U.S.A. (e.g., La Jolla) "would be very pleased to accept the CO_2 and the temperature material, but I do not know whether they would use the latter. I said at least they could have copies of all the CO_2 material."
e. The Historical Group of the Royal Meteorological Society "are interested in preserving historical papers. . . . However, the Met. Soc. Library is very small. . . ."[16]

Anne wrote back (on behalf of Bridget), thanking Schove "for looking after [our] father's papers and notebooks so well." She continued:

They are obviously in good hands. The question is how they can best be made available to the people who need them for further research? My sister and I would of course like his temperature data and his work on CO_2 to be published, but failing that we should certainly wish these sections of his work to go to a research establishment which will make the best use of them. . . . Possibly the best thing to do is for Bridget and I to leave to you the decision as to where the papers should go. I know that we can rely on you to make a good choice. I know that my father would have wished his work to be available to further research, and this is our main concern.[17]

The Callendar papers arrived at the British Meteorological Office in 1984 and were joined by the Schove collection after his death in 1986. They were evaluated by the librarian of the Royal Meteorological Society in 1987 and reached their current destination, the Climatic Research Unit at the University of East Anglia, in 1989.[18] Director of the CRU Tom Wigley wrote to Anne, "We do indeed have your father's papers":

The point of keeping the papers at all is in the hope that someone, either a scientist or a historian of science, will make use of them sometime. As you know, your father was ahead of his time in drawing attention to the greenhouse effect so long ago, and his papers and working documents are an important piece of scientific history.[19]

In 2002, during a visit to the CRU, I noted that the Callendar Papers were not all kept in one location in the library and were not stored properly. Under agreement with the CRU, the collection was loaned to Colby College where they were organized into seven archival boxes, scanned, and returned to the CRU Library.[20] Also in 2002, I met Callendar's daughter Bridget, which led to the accession of an eighth box of family papers by the CRU Library and the completion of the electronic digital archive.[21] Perhaps the collection will continue to grow if others can identify additional relevant material.

The Intellectual Legacy

Callendar counted many, many distinguished scientists as his friends, colleagues, and coworkers. His father was a leading physicist in Great Britain and his first collaborator Alfred Egerton, F.R.S., remained a close colleague throughout his life. Callendar's defense-related efforts, especially his work on the FIDO system that allowed for all-weather flying and saved the lives of many aviators, was, like most of his projects, directed toward non-violent ends. It earned him widespread respect from his peers and admiration from the general public.

His pioneering work on the thermodynamics of steam at temperatures and pressures never before achieved was the foundation of steam-plant design for more than three decades. In 1963 Professor A. G. Gaydon, F.R.S. and Professor of Molecular Spectroscopy at Imperial College, described the Callendar-Egerton collaboration, resulting in publications in the *Philosophical Transactions* in 1932 and 1960, as forming "the basis of the present *International Steam Tables*."[22] H. L. Taylor, writing in 1970, praised the *Callendar Steam Tables* as forming the "bedrock of all steam-plant design calculations in this country for nearly thirty years."[23]

Callendar's climate work has had lasting significance. His 1938 article on climate warming is still widely cited. His 1941 paper on infrared absorption drew rave reviews from meteorologists David Brunt and Sydney Chapman. In 1944, the distinguished British climatologist Gordon Manley made formal note of Callendar's valuable contributions to the study of climatic change.[24] Gilbert Plass, J. Murray Mitchell, and Charles D. Keeling all consulted with

Callendar as they begun their own innovative research programs in the 1950s. Harry Wexler, director of research for the U.S. Weather Bureau, reported on Callendar's results at the 30th anniversary meeting of the American Meteorological Society in 1950:

> CO_2 . . . seems to have a secular trend from .0290 % vol. in 1870–1900 to .0320% in 1935, i.e. an increase of 10% in 50 years, possible effect on heat balance, increased greenhouse effect, warming up (Callendar).[25]

Wexler later supported a program of worldwide CO_2 measurements during the International Geophysical Year.

A decade after Callendar's death, Schove wrote to Bridget: "Your father's work on CO_2 changes is still much quoted in scientific literature. . . . Dr. Callendar's work was discussed recently at a Ph.D. thesis defense. I have also passed some of the reprints to interested scientists, both American and English."[26] Two years later, Ferren MacIntyre at the University of Rhode Island wrote seeking permission to reprint Callendar's 1958 article "On the Amount of Carbon Dioxide in the Atmosphere," in a "definitive collection of papers" on carbon dioxide in the series *Benchmark Papers in Geology*. Although the family granted permission, the volume was never published.[27] In 1985 Tom Wigley, Phil Jones, and Mick Kelly credited Callendar (1961) with making one of the first "reasonably reliable estimates of large scale area average temperatures."[28] And in 2005 Keeling recalled his correspondence with Callendar in the 1950s: "He was a careful investigator and a major contributor to keeping alive interest in the CO_2 Greenhouse Effect during decades when it had almost been forgotten by the science community."[29]

Science-historians, however, have had less to say. One otherwise useful source erroneously claims that Callendar's work was "quickly ignored as World War II intervened and northern hemisphere surface temperatures began to decline in the 1940s."[30] Historian Spencer Weart also errs when he implies that Callendar was a "lone advocate" for the "old idea" represented by Arrhenius's "outmoded hypothesis."[31] Weart is particularly unfair to Callendar when he writes, "Most damaging of all, Callendar's calculations of the greenhouse effect temperature rise ignored much of the real world's physics." First of all, no one who has followed Callendar's technical career in any detail could say this. Weart also misses Callendar's intimate spectroscopic knowledge of the rotational and vibrational absorption bands of trace gases and his studies of band overlap at a realistic effective sky radiation temperature of 263 K. Finally, on his widely cited web site, he mislabels the "only published photo" of Callendar (until now) as coming from a group photograph of the 1929 First International Steam Table Conference in London, when in fact it is from the 1934 American conference (see Figure 3.3).

Brimblecombe and Langford, who worked with the papers, do a much better job in crediting Callendar "with the reinterest in the Greenhouse Effect of the mid 20th century," but portray him in his later years a man with more bitterness to his scientific critics than I think is warranted.[32] I dedicated a section of my 1998 book to Callendar, concerning which Bridget wrote to me in 2002:

> I shall certainly treasure [your book] and only wish Anne could have seen it. She would have been so pleased and proud to read what you have written about Dad. . . . It is so interesting to see [climate change] in its historical perspective, and wonderful to know that Dad is now an authoritative and accepted part of the whole picture. He would have read your book with the greatest interest and been very chuffed, his quiet way, to see his place in it.[33]

In 2005, as mentioned, Mark Bowen referred to Callendar's insights as "pretty much spot-on" and his work on the enhanced greenhouse effect and human agency as being "roughly half a century ahead of his time."[34]

No one until now, however, has undertaken an evaluation of the whole man. In this sense this biography is original. When we recall the strong influence of his famous physicist father and his first-rate education, his association and collaboration with Britain's technical elite on the thermodynamics of steam and the infrared spectra of complex molecules, his defense of his nation applying science and engineering toward nonviolent ends, and the love and support of his wife, daughters, and numerous friends and colleagues, we see a different Callendar, a modest and quiet contributor at the leading edge of research, hitherto unknown, who, among many, many other accomplishments, established the "Callendar Effect," the link between anthropogenic CO_2 and global warming.

Annotated Bibliography of the
Publications of Guy Stewart Callendar

Steam Publications

Callendar, H. L. with G. S. Callendar, 1926. "Direct Measurements of the Total Heat of Steam at Pressures up to 1000 lbs. per sq. inch," *World Power*, 6, 32, 67–76.

"These experiments form a continuation of the investigation undertaken on behalf of the Electrical Research Association. . . . Since measurements of the specific heat at high pressures are exceptionally difficult and liable to error in the neighborhood of saturation, and had not been extended beyond 30 atm. (427 lb.) by Knoblauch and his collaborators, it was decided to continue the direct measurements of the total heat to a limit of 1000 lb. (in the first instance) with a new type of boiler," 67.

Callendar, G. S., 1932a. "The Reduction of Platinum Resistance Thermometers to the International Temperature Scale," *Phil. Mag.* 14, 729–742.

"The following notes describe methods used for the reduction of the observations of resistance change [in platinum resistance thermometers] to gas scale temperature, and are intended to save time in those reductions, and eliminate the use of log tables. Some remarks are also made about the accuracy of individual measurements," 729.

Callendar, G. S., 1932b. *Mollier Diagram—Centigrade Units for Saturated and Superheated Steam*, plotted by G. S. Callendar. London. 8 vo.

Callendar, G. S. and F. E. Hoare, 1933. *Correction Tables for Use with Platinum Resistance Thermometers*. London: Edward Arnold. 12 pp. 8 vo.

Tables with corrections to convert H. L. Callendar's Platinum Thermometer to the International Temperature Scale and corrections with respect to the Standard Value, 2.

Egerton, A. and G. S. Callendar, 1933. "On the Saturation Pressures of Steam (1700 to 3740 C)," *Phil. Trans. Roy. Soc.*, Ser. A, 231, 147–205.

"The apparatus installed at the Imperial College . . . which the late Professor H. L. Callendar devised . . . was easily adapted to the determination of the saturation pressure of steam," 147. "The description is divided into five sections. the first section gives an account of the apparatus, the second the detail of the temperature measurements, the third the detail of the pressure measurements, the fourth the methods of carrying out the experiment and discussion of the possible sources of error, and the fifth the results," 148.

Callendar, G. S. and H. L. Callendar, 1934. *Abridged Callendar Steam Tables. Centigrade Units*, 3rd ed. London: Edward Arnold. 8 pp. 8vo. [4th ed. 1939.] Idem., *Fahrenheit Units*.

Callendar, G. S., Hugh L. Callendar, and Sir A. C. Egerton, 1939. *The 1939 Callendar Steam Tables*. London: Published for the British Electrical and Allied Industries Research Association by Edward Arnold. 66 pp. 4 to. [2nd ed. 1944, 1949, 1957].

Egerton, A. C. and G. S. Callendar, 1939. *The 1939 Heat-Entropy Diagram for Steam, plotted from the 1939 Callendar Steam Tables*. London: Edward Arnold.

Callendar, G. S. and Sir Alfred Egerton, 1960. "An Experimental Study of the Enthalpy of Steam," *Phil. Trans. Roy. Soc.*, Ser. A, 252, 133–164.

This paper is the culmination of the Callendar-Egerton research collaboration between 1930 and 1940. Using a modified form of H. L. Callendar's apparatus, the variation of heat loss with temperature and steam flow was greatly reduced and the variation with cooling effect could be more easily obtained. In 1958 Egerton wrote in his diary, "After the war I was too busy to attempt to wind up the steam work in the form of a paper and the opportunity to do so never arose. But now, particularly in Russia, there is a renewed interest. So this work, done 20 years ago [with G. S. Callendar], is now in the form for general publication" (Egerton, R., Ed., 1963, 61).

Callendar, H. L. and G. S. Callendar, 1962. *Abridged Callendar Steam Tables: CHU version*, 5th ed. Revised by E. J. Le Fevre. London: Edward Arnold. 8 pp., 8 vo.

Callendar, Hugh L., Edwin J. Le Fevre, and G. S. Callendar, 1962. *Abridged Callendar steam tables: Btu version*, 5th ed. London: Edward Arnold. 8 pp., 8 vo.

Defense Reports (Incomplete)

Callendar, G. S., 1942–45. Technical Reports for the Petroleum Warfare Department.

FIDO Project

> Petrol vapour discharge.
>
> A 30 foot model FIDO burner.
>
> A self evaporating liquid fuel burner of the air induction type.
>
> On the use of leaded petrol in evaporating burners.
>
> Trench burners for closing a runway to fog intrusion.
>
> The trench burner.
>
> The distribution of radiation from exposed flames.
>
> Experiments on the thermal melting of ground ice.
>
> The de-icing radiator.

Fuel Propellant Systems and Flame Throwers

> The production of permanent gas at high pressure from a single liquid.
>
> The use of ammonia as a coolant for cordite gas.
>
> The use of dissolved carbon dioxide in pressure accumulators.
>
> Trial of a German flame thrower.
>
> The use of steam for discharging fuel tanks of flame throwers.
>
> The use of petrol vapour for ejecting flame thrower fuel.
>
> Tests with anti surge baffles in a tank.
>
> A vapourized petrol burner for flame thrower ignition.
>
> Pressure losses with fuel in straight pipes.
>
> Experiments on the thermal cutting of wood.

Callendar, G. S., 1950–56. Technical Reports for the Ministry of Supply.

> 1950. "Test of Daniell's 'Dragon' Heater in West Hangar."
>
> 1951. "Trial of a Self Contained Portable Space Heating Unit," Ministry of Supply, A.D.E. Report 3/51.
>
> 1953. "The Diffusion of High Pressure Air into Liquids through Flexible Membranes."
>
> 1956a. "Gravity Method of Obtaining a Low Pressure High Velocity Air Current for Laboratory Research," Ministry of Supply, A.R.D.E. Branch Memorandum S4/8/56.
>
> 1956b. "Preliminary Estimate of Comparative Costs of Four Proposed Methods of Obtaining a Laboratory Scale High Velocity Low Pressure Air Current."
>
> n.d. "Test of Heat Insulated Storage Boxes."

Climate Publications

Callendar, G. S., 1938a. "The Artificial Production of Carbon Dioxide and Its Influence on Temperature," *Quart. J. Roy. Meteor. Soc.*, 64, 223–240.

> [Abstract in *Nature*, 141, 561 and *Meteor. Mag.*, 73 (March), 35. Callendar's annotated copy is in *The Papers of Guy Stewart Callendar*, Digital Edition on DVD,

James Rodger Fleming and Jason Thomas Fleming, Eds. (Boston: American Meteorological Society, 2007), Box 8, Folder 4. Hereafter this work will be abbrevi-
ated as CP.] Discusses links among fuel combustion, rising CO_2 levels, increased
sky radiation, and the observed rise in world temperatures from 200 stations. "By
fuel combustion man has added about 150,000 million tons of carbon dioxide to
the air during the past half century. The author estimates from the best available
data that approximately three quarters of this has remained in the atmosphere.
The radiation absorption coefficients of carbon dioxide and water vapour are used
to show the effect of carbon dioxide on "sky radiation." From this the increase in
mean temperature, due to the artificial production of carbon dioxide, is estimated
to be at the rate of 0.003°C per year at the present time. The temperature observations at 200 meteorological stations are used to show that world temperatures
have actually increased at an average rate of 0.005°C. per year during the past
half century," 223. CP 2, Notebook 1942, 257.

Callendar, G. S., 1938b. "Period Temperature Variations," *Meteor. Mag.* 73 (April),
67–68.

Depicts reversal of the "apparently normal temperature relationship" between
Aberdeen and Beirut, 68. CP 2, Notebook 1942, 256–257.

Callendar, G. S., 1938c. "Note on the Trend of Temperatures, 1880-1937," *Meteor. Mag.*
73 (Aug.), 180–182.

"Movement towards higher temperatures has become especially notable of late
years in North Arctic regions, and also to a less extent in Europe, North America,
Egypt, etc.," 180. CP 2, Notebook 1942, 256–257.

Callendar, G. S., 1938d. "A Century of Temperature Variation in England," *Quart. J.
Roy. Meteor. Soc.* 64, 653–655.

"The longest continuous temperature records observed in Britain show a remarkable consistency throughout at least 98 years to date," 653. CP 2, Notebook 1942,
256–257.

Callendar, G. S., 1939a. "The Composition of the Atmosphere through the Ages,"
Meteor. Mag. 74 (March), 33–39.

"An attempt is made to estimate the most probable changes which our atmosphere
has undergone throughout geological time," 33. "It is a commonplace that man
is able to speed up the processes of Nature, and he has now plunged heavily into
her slow-moving carbon cycle by throwing some 9,000 tons of carbon dioxide
into the air each minute. . . . [T]he best observations show an increase from
0.028 per cent about the year 1900 to 0.030 percent of recent years. . . . As man
is now changing the composition of the atmosphere at a rate which must be very
exceptional on the geological time scale, it is natural to seek for the probable effects of such a change. . . . [T]he principal result of increasing atmospheric carbon
dioxide . . . would be a gradual increase in the mean temperature of the colder

regions of the earth," 38. "Supposed to show the development of the atmosphere, but is very immature, badly written, speculative, with supposed facts tumbling over each other. In fact a real 'student's mag' article. However, some interesting points are made, notably the probable importance of CO_2 in the development of the atmosphere. C. E. P. Brooks said he liked it! 4/61." CP 2, Notebook 1942, 258–259.

Callendar, G. S., 1939b. "Some Interesting Temperature Anomalies for 1938," *Quart. J. Roy. Meteor. Soc.* 65, 137.

"In most North Atlantic countries the mean temperature of 1938 was remarkably high and at several stations in North America, Scotland, and the Baltic countries it equaled or exceeded previous record values," 137. CP 2, Notebook 1942, 258–259.

Callendar, G. S., 1939c. "The Best Climate in the World," *Meteor. Mag.* 74 (August), 208.

"The following figures compare Nelson, [NZ] with the South coast of England," 208. CP 2, Notebook 1942, 258–259.

Callendar, G. S., 1940. "Variations of the Amount of Carbon Dioxide in Different Air Currents," *Quart. J. Roy. Meteor. Soc.* 66, 395–400.

"A brief review is given of the present state of knowledge concerning the variations of the atmospheric carbon dioxide, together with some observations which appear to show that the amount of this gas in the air has increased of late years," 395. "Although the total capacity of the sea water to absorb CO_2 is immense, it is very slow in action, and will doubtless take many centuries to stabilise the great eruption of this gas, now about 300 million cubic metres per hour, which has resulted from human activities," 400. CP 2, Notebook 1942-IRS, 193.

Callendar, G. S., 1941a. "Atmospheric Radiation," *Quart. J. Roy. Meteor. Soc.* 67, 31–32.

Letter regarding W. Elsasser's radiation paper (1940). Elsasser's reply was published in April 1941. CP 2, Notebook 1942, 260. "It seems to be quite commonly assumed by meteorologists, and is of course implied in Elsasser's paper, that the amount of CO_2 in one thickness of the atmosphere can absorb all the energy of which this gas is capable, but there is no observational evidence for this assumption. On the contrary the best measurements show that this amount of CO_2 (2.2 meters), takes up only about 50 percent of the energy in the CO_2 regions. . . . [G]eneral statements regarding the effect of CO_2 in the atmosphere should be based only on observed values of absorption. . . . My own view is that the additional radiation observed at Kew is caused mainly by a local excess of combustion products, such as smoke particles, carbon dioxide, polyatomic gases and vapours which always arise from large centres of population," 32. CP 2, Notebook 1942, 260.

Callendar, G. S., 1941b. "Annual Mean Temperatures," *Quart. J. Roy. Meteor. Soc.* 67, 162.

Letter "Re. Glasspoole's B.I. mean temps in 10/40 Q.J. Suggests the G & H means for the British Isles in 1870 are too high by about 1/40 owing to the use of old type screens at that time." CP 2, Notebook 1942, 261.

Callendar, G. S., 1941c. "Infra-red Absorption by Carbon Dioxide, with Special Reference to Atmospheric Radiation," *Quart. J. Roy. Meteor. Soc.* 67, 263–275.

The mean absorption by CO_2, the effect of pressure on absorption, whole energy absorption, the absorption by mixtures of CO_2 and water vaporand atmospheric transmission of solar radiation through the atmosphere are discussed. "Recent additions to our knowledge of the structure of the water vapour spectrum (Elsasser 1940), and the atmospheric transmission of infra-red radiation (Adel 1939), have tended to emphasise the importance of atmospheric radiation as a fundamental factor in meteorological processes. . . . Normally the greater part of this radiation comes from the large quantities of water vapour present in the air, but there are certain important regions of the atmosphere where the amount of water vapour is extremely small and where a large part of the radiation comes from the carbon dioxide always present. It is probable that measurements of carbon dioxide absorption and radiation have been more numerous and extensive than for most other gases, but these observational data are scattered through the scientific literature of many decades and in several languages. Also they are usually presented in a form which cannot be applied to atmospheric conditions and which requires much coordination and simplification before it can be used for the calculation of energy exchanges. In the following pages these measurements are reviewed and different sets of observations are compared with the aid of a simple function which will give the absorption by any quantity of CO_2 in the different wave bands," 263. "Prof. D. Brunt thought that the paper brought out very clearly the fact that CO_2 absorption was rather more important that had been thought in the past. Most writers had regarded this absorption as limited to a narrow band, so that only a small fraction of the range of wavelengths within the solar band was affected. Mr. Callendar's paper supplied data which would make it possible to estimate with increased accuracy the effects of CO_2 absorption and radiation on the heat balance of the atmosphere," 274. First good summary of the absorptivity (A) of CO_2 much used for references in 1940s and '50s. Revived the old Rubens and Ladenberg observations of A CO_2 as probably still the best for the atmosphere," CP 2, Notebook 1942, 261. "Summary of observed values of CO_2 between 5 and 18." CP 2, Notebook 1942-IRS, 71–75.

Callendar, G. S., 1941d. "The Ocean's Influence on Weather," *Quart. J. Roy. Meteor. Soc.* 67, 383–384.

"I wonder if meteorologists have considered the possibility that recent abnormal conditions in the North Pacific have disturbed the air circulation of our region, so as to give a tendency for sub-normal temperatures in N.W. Europe during the past two years," 383. Letter regarding "warm period in Pacific N.W. 1939–41." CP 2, Notebook 1942, 260.

Callendar, G. S., 1941e. "Climatic Indicators," *Quart. J. Roy. Meteor. Soc.* 67, 385–386.
"In view of the tendency towards milder conditions in the northern regions during the present century, it would be most interesting if a reliable climatic indicator could be found for the southern zone which was comparable in accuracy with the numerous temperature and glacial observations of the north," 385. Letter regarding "rain and temperatures at Santiago, Cape Town, and Adelaide 1861–1938 . . . [s]uggests temp[erature] and precipitation variations at the equatorial margin of the Southern westerlies could be used as climatic indicators. Possible, though doubtful if observations are accurate enough, or period long enough." CP 2, Notebook 1942, 260–261.

Callendar, G. S., 1942. "Air Temperature and the Growth of Glaciers," *Quart. J. Roy. Meteor. Soc.* 68, 57–60.
"There is a close relationship between frontal movements of European glaciers and very small changes in the temperature of preceding years, 58. CP 2, Notebook 1942, 262–263.

Sutherland, G. B. B. M. and G. S. Callendar, 1942–43. "The Infra-red Spectra of Atmospheric Gases other than Water Vapour," *Rep. prog. phys., Lond.* (Phys. Soc.) 9, 118–128 [art. 4].
A critical review, prepared at the request of the Gassiot Committee of the Royal Society, on infrared absorption of gases in the atmosphere. "Confining our attention for the moment to the three well established constituents, H_2O, CO_2, and O_3, it is important to realise that, although all of the major discrete absorption bands can be accounted for qualitatively in terms of the vibration-rotation spectra of these three molecules, we cannot consider the infra-red spectrum as adequately explained until the quantitative side has been fully investigated," 19–20. "A vast amount of work, largely experimental, is required before the problems of radiative equilibrium can be solved. The following seem to us the more urgent requirements: (a) Accurate measurements on the absolute absorption coefficients in the principle bands of CO_2, O_3, and N_2O at a series of different pressures and temperature and in the presence of varying amounts of water vapour. (b) Accurate measurements of the absolute absorption coefficients for at least one band of each of the following: NH_3, CH_4, C_2H_4, C_2H_6, C_3H_8, C_2N_2, HCN, N_2O_4, N_2O_5. (c) Accurate plot of the atmospheric spectrum between 2μ and 7μ, using a slit width of $1\ cm^{-1}$ or less on a grating spectrometer. (d) Estimates of the relative importance

of various absorption bands of the known and the fairly probable atmospheric gases, in the absorption and emission of radiation. This investigation is being started now, even though it cannot be completed until the accurate values of (a) and (b) are available. Such an investigation will, in fact, assist in indicating which gases and which bands should come first in the experimental programme," 27. CP 2, Notebook 1942-IRS.

Callendar, G. S., 1943a. "Temperature Variations in the N.E. Atlantic," *Meteor. Mag.* (May) [Typescript letter].

"Extending Cameron's values for 1930." CP 2, Notebook 1942, 262.

Callendar, G. S., 1943b. "Climate Changes in the North," *Meteor. Mag.* (October– November), 2–3 [Typescript].

"Referring to the interesting article by L. G. Cameron entitled 'The changing temperature of Northern latitudes' in the typescript *Meteorological Magazine*, for March 1941, may I suggest that the charts of temperature variation would be considerably more valuable if a somewhat different change period were used," 2.

Callendar, G. S., 1944a. "Variations of Winter Temperature during Eight Centuries," *Quart. J. Roy. Meteor. Soc.* 70, 221–224.

Re-examination of Easton's (1928) coefficients of winter temperature in France over a period of eight centuries, 222. CP 2, Notebook 1942, 262–263.

Callendar, G. S., 1944b. "Glacial Fluctuations," *Quart. J. Roy. Meteor. Soc.* 70, 231– 232.

"Reply to Seligman's criticism of [Callendar 1942]." CP 2, Notebook 1942, 262.

Callendar, G. S., 1947. "The Climate of the Netherlands," *Quart. J. Roy. Meteor. Soc.* 73 (January), 195–197.

Critical of Labrijin's Dutch temperature reconstructions. CP 2, Notebook 1942, 264–265.

Callendar, G. S., 1948. "Atmospheric Radiation," *Quart. J. Roy. Meteor. Soc.* 74 (January), 81–82.

"It seems to be quite commonly assumed by meteorologists, and is of course implied in Elsasser's paper, that the amount of CO_2 in one thickness of the atmosphere can absorb all the energy of which this gas is capable, but there is no observational evidence for this assumption. On the contrary the best measurements show that this amount of CO_2 (2.2 meters), takes up only about 50 percent of the energy in the CO_2 regions," 82. Commentary on G. D. Robinson's paper in the *Quarterly Journal* for January 1947 on atmospheric radiation. "Establishes that this additional radiation cannot be due to extra CO_2. Proposes mainly dust, smoke, organic vapours, etc., etc." CP 2, Notebook 1942, 265.

Callendar, G. S., 1949. "Can Carbon Dioxide Influence Climate?" *Weather* 4 (October), 310–314.

"Reduced to its simplest terms this theory [of CO_2 climate change] depends on the fact that, whereas carbon dioxide is almost completely transparent to solar radiation, it is partially opaque to the heat which is radiated back to space from the earth. In this way it acts as a heat trap, allowing the temperature near the earth's surface to rise above the level it would attain if there were no carbon dioxide in the air. . . . It is mainly owing to the predominance of water vapour and the extreme irregularity of its absorption that attempts to asses the effect of carbon dioxide on temperatures have so far met with little success. . . . Changes in the amount of carbon dioxide affect the emissivity of the atmosphere at all levels, including the download component at the surface, and these effects can be calculated with some accuracy from the known absorption coefficients," 310. Paper examines regions where the effect of carbon dioxide is greatest, "the possibility of such changes [in CO_2] being brought about by human agency," and the present trend of climate, 311. "It may be said that the climates of the world are behaving in a manner which suggests that slightly more solar heat is being retained in the atmosphere. This could be due to its increasing opacity to terrestrial heat as a result of the additions of carbon dioxide," 314. "A valuable short summary of the probable effect of CO_2 on climate. Most points made were reaffirmed by Hewson and Plass at Toronto Meteorological Conference 1953." CP 2, Notebook 1942, 265.

Callendar, G. S., 1950a."Bleak, Boisterous and Foggy," *Weather* 5 (May), 187.
"Regarding the Climate of Marion Island, 470 S. Latitude," 187. "Suggests Marion as a good site for paleoclimatic studies." CP 2, Notebook 1942, 265.

Callendar, G. S., 1950b. "Note on the Relation Between the Height of the Firn Line and the Dimensions of a Glacier," *J. Glaciology* 1, 8 (October), 459–461.
"There are a number of glaciers which maintain a fairly uniform slope and width for long distances. In such cases it is possible to estimate the effect of a small change in the altitude of the firn line on the length of the glacier, if the change has occurred between periods when it was in climatic equilibrium for several years and was due to a variation of temperature rather than precipitation," 459. "Shows that the length variation of typical glaciers may be 1 to 3 km per degree C. change in temperature of the ablation season." CP 2, Notebook 1942, 267.

Callendar, G. S., 1951a. "The Effect of the Altitude of the Firn Area on a Glacier's Response to Temperature Variations," *J. Glaciology* 1, 10 (October), 573–576.
"The importance of the altitude of the principle accumulation areas of a glacier, when considering its response to temperature changes, has been repeatedly stressed by Ahlmann, Cooper, and others, but the reason for the decisive influence of this altitude is not always very obvious to workers in related fields. It may, however, be clearly demonstrated by means of a simple diagram such as the one accompanying this note," 573. "Shows that temperature variations of ablation

season greatly affect glacier up to 400 m above firn line, slightly 400–700 m, increasingly above 700 m." CP 2, Notebook 1942, 267.

Callendar, G. S., 1951b. "Effect of Shelter Hedges on Mean Temperature," *Meteor. Mag.* 80 (October), 294–295.

"It would appear from these figures that the increasing height and density of the shelter belt at Eskdalemuir has had no significant effect on the overall mean temperature recorded there. Its effect on the daily range is a different matter," 295. "Compares Eskdalemuir, Lancaster, and Blackfort Hill, show no effect of hedge on measurement." CP 2, Notebook 1942, 264–265.

Callendar, G. S., 1952a. "The Greenwich Temperature Record," *Quart. J. Roy. Meteor. Soc.* 78, 265–266.

Greenwich temperatures 1841–1950, with 10, 30, and 50 year curves given. Compares 1925–49 minus 1900–24 for several stations. "A useful note on temperature trends 1840–1950. Greenwich is probably one of the very best European temperature records greater than 100 years." CP 2, Notebook 1942, 264–265.

Callendar, G. S., 1952b. "Air Temperature and Solar Radiation," *J. Glaciology* 2, 11 (March), 69.

Addresses "misconceptions in glaciological literature about the part played by direct solar radiation in the melting of snow and ice," 69. CP 2, Notebook 1942, 267.

Callendar, G. S., 1955. "A Close Parallel Between Temperature Fluctuations in East Canada and Britain," *Quart. J. Roy. Meteor. Soc.* 81 (January), 98–99.

"Compares the 10-year moving average of temperature at Kew with that shown by a group of five stations in southern Ontario," 98. "Finds mean overall rise rate 1870–1950 of 0.020 F/yr in all Canada, 60 stations, and Britain. Useful note on temperature trends, but reference to CO_2 effect in last paragraph is too overcompressed and would be almost totally incomprehensible to most readers of the *Quarterly Journal*." CP 2, Notebook 1942, 266–267.

Callendar, G. S., 1957a. "Contribution to discussion of Plass's paper [vol. 82, p. 310] on atmospheric cooling rate," *Quart. J. Roy. Meteor. Soc.* 83 (April), 273.

"Some useful points are made on the effect of CO_2 in reducing water vapour cooling in lower troposphere. Also on the importance of secondary effects in increasing the influence of CO_2 on climatic change." CP 2, Notebook 1942, 269.

Callendar, G. S., 1957b. "The Effect of Fuel Combustion on the Amount of Carbon Dioxide in the Atmosphere," *Tellus* 9 (August), 421–422.

Letter re: "the large discrepancy between recently measured depressions of the radio-carbon activity, (known as the "Suess effect"), and the apparent increase in atmospheric CO_2 as given by measurements of this quantity between 1870 and 1935," 421. "Suggests possible meteorological and oceanic causes for relatively high CO_2 values obtained by K. Buch in 1930s and notes recent points of interest in new Scandanavian CO_2 [data]. A useful contribution to the heading subject which should be of interest to the USA workers on radio-carbon problems." CP 2, Notebook 1942, 269.

Callendar, G. S., 1957c. "Climatic Changes," *Weather* 12 (February), 67.

"Letter criticizing the diagram used by R. G. Veryard in his article on climatic change in *Weather* 11/56. . . . R. G. V.'s use of accumulated departures from mean of whole period 'effectively obscures any trend.'" CP 2, Notebook 1942, 267.

Callendar, G. S., 1958a. "On the Amount of Carbon Dioxide in the Atmosphere," *Tellus* 10, 243–248.

"Of late years there has been much interest in the effect of human activities on the natural circulation of carbon. This demands a knowledge of the amount of CO_2 in atmosphere both now and in the immediate past. Here the average amount obtained by 30 of the most extensive series of observations between 1866 and 1956 is presented, and the reliability of the 19th century measurements discussed. A base value of 290 ppm. is proposed for the year 1900. Since then the observations show a rising trend which is similar in amount to the addition from fuel combustion. This result is not in accordance with recent radio carbon data, but the reasons for the discrepancy are obscure, and it is concluded that much further observational data is required to clarify this problem. Some old values, showing a remarkable fall of CO_2 in high southern latitudes, are assembled for comparison with the anticipated new measurements, to be taken in this zone during the Geophysical Year," 243. "Continuing 1940 CO_2 paper . . . a useful summary of the observations. Proposes 290 ppm as base value in 1900." CP 2, Notebook 1942, 269.

Callendar, G. S., 1958b. "On the Present Climatic Fluctuation," *Meteor. Mag.* 87 (July), 204–207.

"[Clarification of] the term 'present climatic fluctuation' . . . The figures shown and discussed on the following pages are intended to illustrate the essential difference between the local fluctuations given by decadal averages and long period trends over wide areas," 204. CP 2, Notebook 1942, 269.

Callendar, G. S., 1960. "Contribution to discussion of Kraus's paper on climatic change in vol. 86, Jan.", *Quart. J. Roy. Meteor. Soc.* 86 (October).

"Some useful points made regarding temperatures in the '50s, Simpson's theories, and probable effect of CO_2. Generally more to the point and favorable to Kraus's paper than other contributions." CP 2, Notebook 1942, 269.

Callendar, G. S., 1961a. "Temperature Fluctuations and Trends over the Earth," *Quart. J. Roy. Meteor. Soc.* 87, 1–11.

"The annual temperature deviations at over 400 meteorological stations are combined on a regional basis to give the integrated fluctuations over large areas and zones. These are shown in graphical form, and it is concluded that a solar or atmospheric dust hypothesis is necessary to explain the world-wide fluctuations of a few years duration. An important change in the relationships of the zonal fluctuations has occurred since 1920. The overall temperature trends found from the data are considered in relation to the homogeneity of recording, and also to the evidence of glacial recession in different zones. It is concluded that the rising trend, shown by the instruments during recent decades, is significant from the Arctic to about 45°S lat., but quite small in most regions below 35°N and not yet apparent in some. It is thought that the regional and zonal distribution of recent climatic trends is incompatible with the hypothesis of increased solar heating as the cause. On the other hand, the major features of this distribution are not incompatible with the hypothesis of increased carbon dioxide radiation, if the rate of atmospheric mixing between the hemispheres is a matter of decades rather than years," 1. A manuscript draft copy of this article and the data used in constructing the appendix is in CP 8, Folder 2, Notebook 1960-05-17. In March 1964 Callendar wrote two pages of notes about this article: CP 8, Folder 4, 1961_temperature_note1 and _note2.

Callendar, G. S., 1961b. "Correspondence," Reply to "Temperature Fluctuations and Trends over the Earth," by H. E. Landsberg and J. M. Mitchell, Jr., *Quart. J. Roy. Meteor. Soc.* 87 (July), 435–437.

"I fully agree with their conclusion that the indirect influence on temperature of induced changes in the general circulation may overwhelm primary effects," 436. A copy of this comment and correspondence relating to it is in CP 1. "A valuable summary of the temperature variations over the earth using up to 400 series. An attempt is made for the first time to assess the significance of urban effect's 'improved exposures,' on temperature trends. It is suggested that rising trends are due to increased CO_2 rather than solar heat. Glacial variations are used to substantiate temperature trends." CP 2, Notebook 1942, 271.

Callendar, G. S., 1964. "Climatic Changes Affecting European Birds," *Weather* 19, 264.

Exhange between Callendar and G. Harris regarding the latter's article in *Weather*, March 1964. Callendar emphasized "the necessity for a careful examination of the history of any temperature series before making use of it for generalizations about climatic changes," and cited possible author bias, computational errors, and changes in the location of some stations as reasons for the cooling trend of up to 10C reported by Harris. In reply, Harris pointed out that "Surely Mr. Callendar will agree that the picture of a general decline of temperature in recent years remains unaffected" by these considerations, 264–265.

[P.C.], 1965. Obituary, "Guy Stewart Callendar," *Quart. J. Roy. Meteor. Soc.* 91, 112.

Unpublished Papers on Climate[1]

Note on the Spectrum of Outgoing Radiation from the Troposphere.[2]

Note on Apparatus Required for Measurement of the Infra-Red Absorption of Atmospheric Gases. 1942.

Infra-Red Absorption in Wave Lengths Which Are Important for Atmospheric Radiation.

Simple Empirical Equations for Estimating the Relative Importance of Variables in Atmospheric Radiation. 1948 (Gassiot Commission)

A Contribution to the Problem of Glacial Climates, The Carbon Dioxide Theory. 1953.

The Exchange of Carbon Dioxide Between the Sea and the Atmosphere. 1954.[3]

Book Manuscript

"Climate and Carbon Dioxide"[4]

---

1. Listed in the Red Notebook, 4. CP 8, Folder 5.
2. CP 1. Calledar_n.d._ms_CO_2spectrum.
3. CP 1. Callendar_n.d._charts_oceanCO_2.
4. Not found. See pp. 84–86 this volume.

"Love Letters" from Guy to Phyllis,
September 1934[1]

White Star Line. On Board M.V. "Britannic."

8 Sept. 1934

My dearest Phyllis,

I am safely on board now, after an excellent journey up from home. The only really uncomfortable part was the "Inner circle" from Vic[toria Station] to Euston. I wished I had taken a taxi as it was slow & crowded but I got to Euston at 9:05, & found the Egertons had a reserved carriage for themselves, so there was lots of room for me & bags. We had sandwich lunch & coffee at about 12:00. Very comfortable smooth train, with large low windows like a Pullman. It was weird going through the docks on the train as we seemed to get lost in miles of warehouses, which were full of packages, & seemed to have only tramlines going through them, however we got to the ship in the end at 1:45. It is a fine boat with lots of lounges, smoke rooms, etc., etc. & does not seem crowded as there are such hundreds of armchair in the various rooms. Some of the latter are very fine with huge imitation log fires, lots of alcoves, fine wood paneling, & so on. I will tell you more about it later on.

My cabin is very nice, you would love it, several little cupboards, basin,

---

1. CP 8, Folder 1. See Chapter 3 in this volume.

[hot and cold water], air blows in through a shutter, and the bed looks very comfortable, also an armchair in it & reading light.

I hope you had a good rest in bed after I left. I wonder if the twins missed me today, you must show them the picture of the boat & tell them I'm there.

You would like the tea here, just sit down in an armchair, & have it all brought to you without asking, cakes, muffins, etc. My cabin is about the size of my little room at home.

I hear the Galway mail goes out on Sunday night, so I will send this letter then. It is lovely & sunny this evening & the sea is very blue with a strong breeze. The ship seems quite steady in these small waves, but there is a slight vibration from the engines. We have eight at our supper table: the Egertons, Guys, Robinsons, myself & another man I don't know yet. Not dressing tonight, but will be all other nights.

Sunday.
Lovely morning in Cork harbour, thank you ever so much for your telegram, and the nice little note you left in my bag, it was sweet of you to think of them.

Last evening we had a cinema show in the large lounge, it was some frightful murder thriller. I sympathized with the Egertons who I should think had never seen anything like it!

Supper was the usual 5-course affair; very good food and lots of things to choose from. I slept quite well, though the air makes rather rushing noise, as it comes into the cabin, the water was hot enough to shave with! There does not seem to be any tennis court, but two deck tennis courts instead.

We have great arguments at meals on the Soviets & other political subjects, that is the men do, the ladies talk of more personal things. I have not seen much of the Egertons since we came aboard, but a lot of the Scotch engineer and Mr. Guy. We are running along the south coast of Ireland this morning with blue sky and sea with some rollers on it, but the ship is not moving much to them.

Later.
We went round the south west corner of Ireland this afternoon, the coast is very fine, there lots of islands & headlands, etc.

We spent this evening having a miniature steam conference, so it has got very near to posting time for Galway. I have tried all the games on board & like the deck tennis best, it is just like badminton except that you catch a ring, instead of hitting a shuttle.

The boat is still steady, but there is a depression coming up, and I expect it will be rougher tomorrow. It looks like a mild low to me, so I expect it will be more rain than wind.

Well darling I must say goodbye now. I do hope the twins are not being too troublesome, and that you are not very lonely at night. I showed the Egertons the photos & they thought them lovely. Mrs. asked me to send you her best regards. I must stop at once or I shall miss the post.

Lots of Love darling.

From your own Guy. XXXxxxxxx

White Star Line. On Board M.V. "Britannic."

Monday 10th Sept.

My dearest Phyllis.

What a contrast, yesterday we were sitting in the sun watching the green hills and rocky islets of the Irish coast go by, this morning the rushing wind is everywhere with flying spray. The log reads "Mod to fresh s.w. gale, heavy seas sending spray over the poop." The boat is pitching in a dignified way, the front part falling & rising as much as 50 ft. at times (27 [ft.] herald[2] is about 25 [ft.] to roof line), about 90% of the passengers are more or less queer, more than 50% keeping to their cabins. I am feeling moderate, but not able to face the dining room meals so I am subsisting on chicken sandwiches from the buffet. I have lovely deep sea water baths, with a fresh water shower to follow.

Tuesday.

Had a good sleep & feel better, had breakfast in the saloon. This is queer weather, with bright sun most of the time, and a tearing wind from southwest, sea rather smoother than yesterday, but still equal to worst winter gale at Hastings, and of course much higher than anything you would see at Worthing. Later on the wind dropped & it rained a little, many fresh people appeared on deck.

---

2. Nautical term for plate on the stern of a ship inscribed with the ship's name.

Quite a good cinema show tonight, but too full of close ups showing lush kisses. Dancing in the saloon 9 to 12 pm.

You would like the meals here today. I had supper as follows:

Bronx cocktail[3]
1/2 cantaloupe melon
Red Mullet
Surrey Capon
Garden peas & asparagus
Large strawberry ice on pineapple chips
Fresh peach
Coffee.

Of course there are innumerable other choices. The melons to start & ices to finish are nice, otherwise I don't really enjoy these elaborate meals, more than ours at home.

Wednesday.
Cool & fresh with sun & cold wind this morning, they have an illuminated map showing the position of the ship, we are south of you now & 1/2 way across. I get plenty of exercise walking round & playing the various deck games.

We hold small steam conferences which last most of the morning to explain the various reports.

Thursday.
Had a fancy dress carnival last night, tremendous amount of noise with toy trumpets, balloons, etc. It beats me how grown up people can find amusement blowing tin trumpets, of course some of them had had the necessary amount of alcohol, but some of our party had not had any! Mr. & Mrs. Robinson were very good as Egyptian nobles. The Secretary of the Royal Society, Dr ——, stood us cocktails before supper, of course one cocktail is nothing, not equal to a pint of beer, though it may cost twice as much.

Fair with cold wind this morning, sea still rather rough, but most of the passengers are all right again now.

---

3. Made with vermouth, gin, orange juice, and orange slice.

I am missing you and the babies quite a lot, it is fortunate there is so much to do. Mrs. Egerton say[s] you would enjoy this trip far more·than I do, and I'm sure she is right, though I am enjoying it very well, I am astonished what small things seem to amuse people on board, I suppose very few hold anything in their head which interests them.

Friday.
Lovely morning sun & calm sea. We are running along the south coast of Newfoundland now, we sighted Cape Race at 7:00, it looks a rocky and desolate coast compared with Ireland, though it is hundreds of miles further south than the latter. The Newfoundland winter is far more severe than anything we get in the British Isles.

Had a cinema last night, "Judge Priest," turgid with emotion but of a much higher class than most films. I had a dream about you last night, I dreamt I was looking over the ship for you & a[t] last found you watching the film, I felt pleased all over when I found you and wanted to give you lots of real kisses, but woke up instead feeling frightfully homesick, fortunately I went to sleep again fairly soon.

I am sticking to white meats at the meals as I think they are less likely to upset one, the ices are extremely good, very nearly equal to those at Tregenna.[4]

I have not spent anything more than a few shillings yet, as there are no extras except beer, of course 801 per day ought to cover most things![5]

Saturday.
Fine again with blue sea, we are off Nova Scotia this morning but cannot see land yet.

Went over the engines of this ship with Mr. Guy yesterday, they are i.c. [internal combustion] that is like those in a motorcar. The engine room is like 500 large lorry engines all going together in a tin shed, it is no place for ladies.

Your packing was very good dear, the only thing missing was that blue packet of razor blades, but I got some more at the shop here so it did not matter at all. We are supposed to get in to Boston at 10 AM Sunday, I shall

---

4. Possibly a reference to Tregenna Castle Hotel, St. Ives, Cornwall.

5. £8, 1 p per day.

post this on board so as to use my English stamps. I will write again in a few days, but not so long as this because the conference will be on then. I shall not dress for dinner tonight as my evening suit will be packed up then.

Great excitement this afternoon as we ran through a school of w[h]ales. I must rush off & pack now as the bags must be ready by 9:00.

Lots of Love to you darling.

From your own Guy.

Love and kisses also for the twins, I am looking forward to hearing all about them in your letter.

Hotel Astor, New York
20 & 21 Sept. 1934
My dearest Phyllis,

Thank you so much for your lovely letter, which came in last evening, you can't think how much I enjoyed reading it. Anne's remarks made me feel frightfully homesick.

I have been rushing round so much lately that I have not had a minute to write.

It is lovely this morning with a bright sun shining on the high New York buildings.

I will give a brief account of what we have done since my last letter. We got into Boston early on Sunday morning in a fog, it took 2 hours to get through the customs. Some American professors met us & took us to lunch at Harvard. In the afternoon they took us for a nice motor ride to see the local points of interest, New Inn [?], Concord, etc., etc. Nice residential districts round Boston, which has a magnificent harbour with small islands in it.

Train to Washington 12 hours in rather uncomfortable sleeper with common wash place. Fine city with solid looking white buildings.

Back to Boston on Tuesday to see the M.I.T. apparatus & hear interesting lectures. By very comfortable boat to New York for the main conference.

My room at the Astor is very nice with comfortable armchairs, & private bathroom attached. Last evening we went to the top of the Empire State Building 1300 ft high (200 above the Eiffel Tower) & looked down on all the lights of New York.

We have dinner every night, on Wednesday 7 to 11:30, speeches nearly sent everyone to sleep. The result is that I have not a moment to write, I should be having breakfast now! I shall have to tell you about the things I have seen, when I get back, as it would take 10 pages to convey any idea of them. How sweet of you to say you are going to try not to be cross with me ever when I get back, I think only an angel could do that with two troublesome twins to look after, but I know you will do your best darling, not to let little things bother you. Try & leave them occasionally with the family, if they will let you go.

All the ladies of our party have seen the photos of them & they are greatly admired.

Yes certainly you can go up to town one day when I get back & I will give you your fare. Thank you for the article by Armstrong.

Lots of kisses to you & the babies

from

Your ever loving

Guy.,

Hotel Astor, New York

22 Sept. 1934

My dearest Phyllis,

Not quite so busy today as the conference is nearly over, it is voted a tremendous success by all concerned. We had a supper last night at the Waldorf-Astoria which is a new, and very super hotel, I must say I found the chromium plate & polished wood interiors very pleasing. This supper was given by the visiting delegation, Mrs. Egerton made a speech for the ladies and sent all their best wishes to you and the twins, she mentioned the latter by name & said that they were the "darlingest little girls you ever saw." Mrs. Key[e]s, the little French wife of Prof Keyes, had previously sent the greetings of the American ladies to you & the twins. They were also mentioned in the course of a humorous speech by Dr. Davis, president of the American society of engineers.

On Wednesday we were entertained to dinner at the N.Y. Society of Arts, by Mr. Orrich, & went on a sightseeing expedition afterwards. I may play tennis with Keenan and Mrs. on Sunday afternoon.

I am looking forward to a quiet time during the next four days, as apart from the above & a visit to Philadelphia on Wednesday next, I am not fixed up.

I learn that I can wear grey trousers for the tennis & that Keenan can lend me a racket. I have purchased a magnificent pair of tennis shoes for 6/— they would cost 18 to 30/— in England.

Now that all the rush and work of the conference is over I feel frightfully homesick, and long to be back at our quiet little home where I really belong. Once the boat starts moving I shall be all right, I see the Britannic started back this morning, I wish I was on her. Many people would think I must be mad to wish to quit living like a lord, free of charge, in the center of New York, but I want to get home to you and the twins. The air here is like warm thick soup, there is no movement, and it tastes & smells *very* secondhand. I have a huge fan in my room, but the air in the streets (canyons) is awful. I long for the glorious fresh breezes of Worthing.

I have a little green disc set in the wall of my room (No 676) which lights up if a message for me is at the office, it is like a link with you because the only time it has lit was for your letter. A fast German boat leaves on the 26[th] & gets to Southampton on the 1st, but unfortunately I cannot transfer to her.

Some reaction is inevitable after the excitement of the past week. & I shall feel right again tomorrow especially after a game of tennis. I wish I had brought a photo of you as well as the babies, Mrs. Key[e]s who is a little thing like you, asked to see one of you as well as the twins & I was a fool not to have one to put in my room.

Well darling I will post this now so as to get to you as soon as possible. I will write another in a day or two.

All my love & kisses to you and the babies.

Your own devoted

Guy.

Monday 24 Sept. 1934

My dearest Phyllis.

I am told I must post a letter not later than tomorrow if I wish it to reach you before I do.

I have had a good rest since the conference closed, I have been for one or two trips on charabancs[6] to see New York, but until today the weather has been so hot & close that I have kept pretty quiet. Today it has been fine & sunny and I had quite a good game of tennis over at Hoboken with two

---

6. A large open-topped sightseeing bus.

of Keenan's friends (and Keenan himself). From the court we had a view of the famous New York skyline, on the other side of the Hudson River. The students at this university play tennis with only the shortest of shorts on, and no shirt, which we should think hardly decent as there are plenty of girls about. Tomorrow I am taking the bus to Phyladelphia [sic.] to meet the Egertons who are at present in Washington, and we are to be shown round on Wednesday by our friend Lascells whom we met on the boat. I return here on Wednesday eve or Thursday morning.

I am rather troubled as to what present to get you, as the handbags seem very ordinary, as do most other things in the shops. I will find something dear, though it may be quite small. I am longing for the time when the ship actually starts to move towards home, I am going down to the shipping office tomorrow morning to find out what time of day it starts.

Tuesday morning.
I see from the paper that the Laconia has bumped a cargo vessel in a fog, it says she is not damaged except for scratches, I am just going down to the office to enquire. I trust her sailing date will not be delayed. I have brought a very comprehensive book on combustion, Keenan got it for me at 20% discount.

I am seeing if there is a chance of booking on the Mauritania, sails tomorrow, if so I shall be home before this.

In haste.
Lots of Love
from
your own
Guy.

Inventory of G. S. Callendar's
Study at the Time of His Death

This inventory was compiled in October 1964 by Callendar's daughters, and is transcribed here with minor emendations from the Red Notebook. CP 8, Folder 5.

Technical Reference Books

Oceanography for Meteorologists. H. V. Sverdrup. 1945.

Cambridge Five-Figure Tables. Hall and Rideal. 1929.

Variations of Temp. at Oxford 1815–1934. L. F. Lewis. 1937.

Enlarged Callendar Steam Tables. 1924.

The 1939 Callendar Steam Tables.

Glaciers and Climate. H. W. Ahlmann. 1949.

Chambers Mathematical Tables (circa 1900).

Revised Steam Tables and Diagrams of the J.S.M.E. by the Japan Soc. of Mech. Engineers. 1950 and 1955.

Thermodynamic Properties of Steam. Keenan and Keyes. 1936.

Physical and Chemical Constants and Some Mathematical Functions. Kaye and Laby. 1920.

Fuels and Their Combustion. Haslam and Russell. 1926.

Reports on Progress in Physics. Vol. IX, 1942–43.

Newnes Engineers Manual. F. J. Camm. 1941.

Thermodynamics for Engineers. Ewing. 1936.

Text Book of Inorganic Chemistry, Vol. V. Caven. 1921.

Coal: Its Constitution and Uses. Bone and Himus. 1936.

Coal and Its Scientific Uses. W. A. Bone. 1919.

Hints to Meteorological Observers. Marriot. 1887.

The Meteorology of the Falkland Islands and Dependencies. J. Pepper. 1944–1950.

Copies of Papers published by G. S. C.

Abridged Callendar Steam Tables—Centigrade.

Abridged Callendar Steam Tables—Fahrenheit.

The Reduction of Platinum Resistance Thermometers to the International Temp. Scale. (Phil. Mag., November 1932).

Correction Tables for use with Platinum Resistance Thermometers. Callendar and Hoare. 1933.

An Experimental Study of the Enthalpy of Steam. Callendar and Edgerton. 1960. (Roy. Soc.)

On the Saturation Pressures of Steam. Edgerton and Callendar. 1932 (Roy. Soc.).

The Effect of the Attitude of the Firn Area on a Glacier's Response to Temperature Variations. (Journal of Glaciology, October 1951).

Note on the Relation Between the Height of the Firn Line and the Dimensions of a Glacier. (Journal of Glaciology. February 1950).

Note on the Effect of Temperature on Glaciers. 1941.

The Composition of the Atmosphere Through the Ages. (Meteorological Mag. March 1939).

On the Present Climatic Fluctuation. (Met. Mag. 1958).

Variations of Winter Temperature During Eight Centuries. (Quarterly Journal. July 1944).

A Close Parallel Between Temperature Fluctuations in East Canada and Britain. (Q.J. Jan. 1955).

The Greenwich Temperature Record. (Q.J. April 1952).

Temperature Fluctuations and Trends Over the Earth. (Quart. Journ. January 1961).

Variations of the Amount of Carbon Dioxide in Different Air Currents. (Q.J. October 1940).

Infra-Red Absorption by Carbon Dioxide with Special Reference to Atmospheric Radiation. (Quart. Journ. July 1941).

The Artificial Production of Carbon Dioxide and Its Influence on Temperature. (Q.J. April 1938).

Can Carbon Dioxide Influence Climate? (Weather October 1949).

The Effect of Fuel Combustion on the Amount of CO_2 in the Atmosphere. 1957 (Letter in Tellus).

On the Amount of Carbon Dioxide in the Atmosphere. (Letter in Tellus 1957).

On the Amount of Carbon Dioxide in the Atmosphere. (Tellus 1958).

A Century of Temperature Variation in England. October 1938.

Annual Mean Temperatures. April 1941.

The Ocean's Influence on Weather. October 1941.

Climatic Indicators. October 1941.

Atmospheric Radiation. January 1941.

Air Temperature and the Growth of Glaciers. January 1942.

Air Temperature and Solar Radiation (Letter in Journal of Glaciology 1951).

Unpublished Papers

Note on the Spectrum of Outgoing Radiation from the Troposphere.[1]

Note on Apparatus Required for Measurement of the Infra-Red Absorption of Atmospheric Gases. 1942.

Infra-Red Absorption in Wave Lengths Which Are Important for Atmospheric Radiation.

Simple Empirical Equations for Estimating the Relative Importance of Variables in Atmospheric Radiation. 1948 (Gassiot Commission).

A Contribution to the Problem of Glacial Climates, The Carbon Dioxide Theory. 1953.

The Exchange of Carbon Dioxide Between the Sea and the Atmosphere. 1954.[2]

Reprints of Papers by Other Meteorologists

Abbot. Sun Spots and Weather (1933).

Ahlmann. Glaciological Research on the North Atlantic Coasts (Royal Geog. Society, 1948).

Arakawa. Selected Papers on Climatic Change (Met. Research Inst., Tokyo, Japan).

Barnes. The Crystal Structure of Ice Between 0°C and −183°C (Royal Society, 1929).

Bonacina and Hawkes. Climatic Change and the Retreat of Glaciers (1947).

Brinkworth. The Ratios of the Specific Heats of Nitrogen at Atmospheric Pressure and at Temperatures between 10°C and −183°C (Royal Society, 1926).

---

1. CP 1, Calledar_n.d._ms_CO_2spectrum.
2. CP 1, Callendar_n.d._charts_oceanCO_2

Burch, Gryrnak and Williams. Infrared Absorption by Carbon Dioxide. (Ohio State Univ. Research Foundation. January 1961).

Cowling, T. G. Atmospheric Absorption of Heat Radiation by Water Vapour.

Cowling, T. G. The Calculation of Radiative Temperature Changes (1950).

Craig. Tellus (1956).

Dines. Monthly Mean Values of Radiation from Various Parts of the Sky at Benson, Oxfordshire (Royal Met. Soc., 1927).

Drummond. Cold Winters at Kew Observatory (Royal Met. Soc., 1942).

Eggers and Crawford. microfilm of paper on CO_2 (1951).

Fairbridge, R. W. New Radiocarbon Dates of Nile Sediments (Nature, 1962).

Flohn. Meridional Transport of Particles and Standard Vector Deviation of Upper Winds (1961).

Flohn. Man's Activity as a Factor in Climatic Change (New York Academy of S., 1961).

Glueckauf (1951).

Heusser, C. J. Variations of Blue, Hoh, and White Glaciers During Recent Centuries (Arctic, 1957).

Hills. The Cooling of the Earth by Convection and Some of Its Consequences (Empire Survey Review, 1939).

Hustich. Correlation of Tree-Ring Chronologies of Alaska, Labrador, and Northern Europe (Helsinki, 1956).

Kaplan, L. D. The Influence of Carbon Dioxide Variations on the Atmospheric Heat Balance (Tellus, 1960).

Kaplan and Eggers (1956).

Lamb, H. H. Atmospheric Circulation, Climate, and Climatic Variations (Geography, 1961).

Lamb, H. H. Halley Bay Meteorology (1960).

Lamb, H. H. Climatic Change Within Historical Time as Seen in Circulation Maps and Diagrams (1961).

Lamb, H. H. and Johnson. Climatic Variation and Observed Changes in the General Circulation (1961).

Landsberg and Mitchell. Temperature Fluctuations and Trends Over the Earth (Correspondence, Quart. Journ. July 1961).

Lane and Urey. Ages by the Helium Method: I. Keweenawan (Geological Soc., 1934).

Ludlam (1957).

Manley, Gordon. The Snowline in Britain (1949).

Manley, Gordon. Topographical Features and the Climate of Britain (1944).

Manley, Gordon. Some Recent Contributions to the Study of Climate Change (1944).

Manley, Gordon. The Range of Variation of the British Climate (Geog. Journal, 1951).

Mitchell, J. M. The Measurement of Secular Temperature Change in the Eastern United States (Office of Climatology, U.S. Weather Bureau, 1961)

Plass. The Influence of Infrared Absorptive Molecules on the Climate (1961).

Plass. The Influence of Numerous Low-Intensity Spectral Lines on Band Absorption (1964).

Plass. Transmittance of Carbon Dioxide and Water Vapour over Stratospheric Slant Paths (Applied Optics, 1964).

Plass. Spectral Band Absorptance for Atmospheric Slant Paths (Applied Optics, 1963).

Plass. Mie Scattering and Absorption Cross Sections for Aluminum Oxide and Magnesium Oxide (Applied Optics, 1964).

Plass and Stull. Carbon Dioxide Absorption for Path Lengths Applicable to the Atmosphere of Venus. Technical Report, Ford Motor Co., USA.

Plass, Stull, and Wyatt. Final Report on Theoretical Study of the Infrared Radiative Behavior of Flames. Vol. I: The Infrared Absorption of Water Vapour. Vol. II: The Infrared Absorption of Carbon Dioxide, 1961. (Prepared for the Geophysics Research Directorate, United States Air Force, Mass.).

Plass, Stull, and Wyatt. Infrared Transmission Studies. Vol. I. Spectral Band Absorptance for Atmospheric Slant Paths (1962); Vol. II. The Infrared Absorption of Water Vapour (1962); Vol. III. The Infrared Absorption of Carbon Dioxide (1963); Vol. IV. The Influence of Numerous Weak Lines on the Absorptance of a Spectral Band (1963); Vol. V. Transmission Tables for Slant Paths in the Stratosphere (1963); Vol. VI. Mie Scattering and Absorption Cross Sections of Aluminum Oxide and Magnesium Oxide (1963).

Plass, Stull, and Wyatt. Quasi-Random Model of Band Absorption (J. Optical Soc. Amer., 1962).

Plass, Stull, and Wyatt. The Infrared Transmittance of Carbon Dioxide (Applied Optics, 1964).

Plass, Stull, and Wyatt. The Infrared Transmittance of Water Vapour (Applied Optics, 1964).

Plass, Stull, and Wyatt. Vibrational Energies of the CO_2 Molecule (J. Chem. Phys., 1962).

Revelle and Suess (1956).

Riehl (1956).

Robinson (1947, 1949).

Rodenwald (n.d.).

Schell. The Ice Off Iceland and the Climates During the Last 1200 Years Approximately (1961).

Schell. Recent Evidence About the Nature of Climate Changes and Its Implications (1961).

Schove, D. J. Chinese 'Raininess' Through the Centuries (Meteorological Mag., January 1949).

Schove, D. J. The Reduction of Annual Winds in North Western Europe, A.D. 1935–1960.

Schove, D. J. Summer Temperatures and Tree-Rings in North Scandinavia, A.D. 1461–1950.

Schove, D. J. Models of the Southern Oscillation in the 300-100 mb Layer and the Basis of Seasonal Forecasting (1963).

Schove. D. J. The Major Pressure Oscillation 1875 to 1960 (1961).

Schove, D. J. Auroral Number Since 500 B.C. (1961).

Smed (1949).

Symons. On Barometic Oscillations During Thunderstorms, and on the Brontometer, an Instrument Designed to Facilitate Their Study (1890).

Taylor, Benedict and Strong. microfilm of paper on CO_2 (1952).

Thorndike. microfilm of paper on CO_2 (1947).

UNESCO Rome Symposium in Paris 1961. Papers presented on changes of climate with special reference to the arid zone.

Wiseman and Ovey. Recent Investigations of the Deep Sea Floor (Geologists Assn., 1950).

Magazines—English Societies

Journal of Glaciology (25 magazines, various dates between January 1947 and February 1964).

Quarterly Journal of the Royal Meteorological Society (various dates between 1938 and 1964).

Ice (few copies).

Proceedings of the Royal Society, Series A, vol. 126, January 1930.

Meteorological Office: Averages of Temperatures for the Brit. Isles, periods ending 1935.

Meteorological Translations of Papers on Climate (Dept. of Transport, Toronto).

The Meteorological Magazine (copies from 1938–1959).

Weather (copies from 1946–1964).

Monthly Weather Report of the Meteorological Office (covering many years).

Centenary Proceedings of the Royal Met. Society, 1950.

Proceedings of the Toronto Meteorological Conference, 1953.

Geographical Journal. March 1951.

Royal Met. Society. Recent Additions to the Library. 1961–1964.

University of California—Scripps Institution of Oceanography. Progress Reports (1957).

Royal Society. Gassiot Committee. Reports.

Various Forestry Journals, including:

Quarterly Journal of Forestry

Reports on Forest Research

Forestry Commission booklets

Forest Products Research

Magazines—Foreign Societies

Annals of the New York Academy of Sciences, vol. 45 [or 95?]. Solar Variations, Climatic Change and Related Geophysical Problems.

Australian National Antarctic Research Expeditions. Interim Report 7. Heard Island.

Glaciologie Groenland par A. Bauer.

Daily Observations in Wroctaw (various dates 1947–1955).

Polish I.G.Y. Spitsbergen Expeditions in 1957, 1958, 1959. (Wroctaw).

Institut Royal Météorologique de Belgique.

L'Isola D'Ischia Climatica.

WMO Bibliography on Climatic Fluctuations (World Met. Organization).

Papers by Dr. F. Prohaska, Dr. H.U [or V.]. Roll [or Rau], M. Schiiepp, M. Nicolet.

G. S. Callendar's Notebooks

(About 100 notebooks, compiled mainly during the last 10 years. Some earlier. See CP for full details.)

Majority of notebooks contain climatic data and temperature series from weather stations all over the world. (Temperature figures mainly of 19th and 20th centuries—some from 18th). Some books devoted specifically to stations in one country. Others compare data from different countries, e.g.

Regional and Zonal Summaries

Deviations of Temperature from Average (monthly, annual, 5 yr., 10 yr., 30 yr., 50 yr., etc.)

Seasonal Variations

Moving Average Temperature Curves

World Weather Records

WMO Station Lists

Selected Temperature Series—the 100 best stations

Additional Stations, 10 yr. temperature means

British Isles Records—period means etc.

Comparison of Stations in British Isles

Daily Records ("rough") of "Percuil" weather at 44 Parsonage Road, Horsham, and comparisons.

Notebooks containing notes and references of work by other meteorologists. Comparison of data.

Notebooks containing notes and figures for own papers (e.g. "Temperature Fluctuation," etc.).

Notebooks specifically on "CO_2 and Climate" data.

Notebooks containing notes and work on glaciers.

Notebooks containing data from the Meteorological Office Library.

Data on specific subjects, e.g.

> Sea Temperatures
> Sky Radiation
> Radiation Losses from a Surface Partially Screened
> Climatic Indices
> Lake Freezing.

Instruments (Imperial College?)

Calculator (in wooden box)

Set of Mechanical Drawing Instruments for Engineers (in box)

Metal device for drawing accurate convex lines

Micrometer Caliper (in box)

Vernier gauge (in box)

Try and mitre square with spirit level

Small brass microscope (without stand)

Celluloid measuring instruments, including Set Square for Callendar Steam Diagram.

Correspondence

Letters on climate and temperature etc. exchanged with the following:

Allmann, Arakawa, Barnett (N. Zealand Met. Service), Brooks, Davy (Mauritius), Dixon, Flohn, Harris, Heusser, Kraus (Colorado), Kosiba, Keeling (and Plass), Kaplan (and Sutherland), Lamb, Longley, Manley, Mitchell, Plass, Roberts (Scott Polar Research), Rubenstein (Russia), Rodenwald, Schove, Schell. Also letters to Royal Met. Soc. About "Temperature Fluctuations."

(The correspondence contains rough copies only of Callendar's replies.)

Temperature and Climate Data on Loose Paper. G. S. C.
5 year Dv.

Urban and Rural Sites (Temperatures).

Curves—CO_2—Sea.

Zonal Curves—3 yr., etc.

Kew Temperatures. Seasonal Comparisons, etc.

Figures and Plates for "Climate and Carbon Dioxide".

A/CO_2 H. B. and W. 1956.

Curves.

A/log CO_2 curves.

British Isles. Annual Temperatures.

A/Co2 Curves. Sky Radiation Curves. CO_2 Temperature, etc.

List of Physics Books not given to Schove
Revised Steam Tables and Diagrams of the J.S.M.E. 1955.

Revised Steam Tables and Diagrams of the J.S.M.E. 1950.

Thermodynamic Properties of Steam. Keenan and Keyes. 1936.

Fuels and Their Combustion. Haslem and Russell. 1926.

Coal: Its Constitution and Uses. Bone and Himus. 1936.

Physical and Chemical Constants. Kaye and Laby. 1920.

Newnes Engineers Manual. F. J. Camm. 1941.

Cambridge Five-Figure Tables. Hall and Rideal. 1929.

Coal and Its Scientific Uses. W. A. Bone. 1919.

Text Book of Inorganic Chemistry, Vol. V. Caven. 1921.

Keep
Reports on Progress in Physics. Vol. IX, 1943.

Thermodynamics for Engineers. Ewing. 1936.

Annals of the New York Academy of Sciences, vol. 95, Art. 1.

The Enlarged Callendar Steam Tables. 1924.

Ice (few copies).

Proceedings of the Royal Society, Series A, vol. 126, January 1930.

The 1939 Callendar Steam Tables.

Notes

Abbreviation

CP refers to *The Papers of Guy Stewart Callendar*, Digital Edition on DVD, James Rodger Fleming and Jason Thomas Fleming, Eds. (Boston: American Meteorological Society, 2007).

Introduction

1. A brief account of Callendar's work on climate change appeared in James Rodger Fleming, *Historical Perspectives on Climate Change* (New York: Oxford University Press, 1998, 2005), vii and 113–118.

Chapter One

1. In Chapter 1, "Hugh" and "Callendar" refer to H. L. Callendar; throughout the book "Guy" refers to G. S. Callendar. In later chapters, "Callendar" refers to G. S. Callendar. Standard sources for H. L. Callendar's life include J. R. H. Weaver in *Dictionary of National. Biography 1922–1930* (Oxford: Oxford University Press, 1937), 152–154; *Who Was Who*, Vol. III, *1929–1940* (London: Adam & Charles Black, 1941), 209; "Obituary Notice of Hugh Longbourne Callendar," *Proc. Roy. Soc.* A (1932): 134; and *Encyclopedia Britannica Online*, http://www.search.eb.com/eb/article?eu=18989 (21 September 2003). Documentation of H. L. Callendar's birth

is in IGI (International Genealogical Index), Batch no. C029042, Dates 1838–1876, Source: 855633 Film, 6901837 Fiche.

2. L. H. Callendar, "H. L. Callendar—Instrument Engineer," *The Chartered Mechanical Engineer*, London (Feb 1966): 67–72; copy in CP 8, Folder 8.

3. L. H. Callendar, "Professor H. L. Callendar," *Bulletin of the Institute of Physics and the Physical Society*, London (April, 1961): 87–90; copy in CP 8, Folder 8.

4. L. H. Callendar, "H. L. Callendar—Instrument Engineer."

5. Books by H. L. Callendar on shorthand include *A Primer of Cursive Shorthand: The Cambridge System* (1889); *A Manual of Cursive Shorthand* (1889); *Reading Practice in Cursive Shorthand* (1889); and *A Manual of Orthographic Cursive Shorthand: The Cambridge System* (1891); all published in London by C. J. Clay.

6. "J. J. Thomson—Biography," http://www.nobel.se/physics/laureates/1906/thomson-bio.html (23 March 2004). Thomson had recently won the Adams Prize for his *Treatise on the Motion of Vortex Rings*.

7. The instrument was an improvement of an earlier device created by Sir William Siemens; see London Science Museum, *Temperature Measurement and Control Catalogue 78*, Inventory 1930-360. J. J. Thomson, quoted in L. H. Callendar, "Professor H. L. Callendar."

8. H. L. Callendar, "On the Practical Measurement of Temperature: Experiments made at the Cavendish Laboratory, Cambridge," *Phil. Trans.* A, 178 (1887): 161–230.

9. For details on early and later model of Callendar thermometers, see London Science Museum, *Catalogue of the Collections in the Science Museum, Temperature Measurement and Control*, Part II, by J. A. Chaldecott (London, 1955, 1976), Inventory 1930-363.

10. Ibid.

11. London Science Museum, Inventory Card, 1899-55.

12. Ibid., 1930-356.

13. Ibid., 1930-349.

14. Ibid., 1930-347; 1930-348; 1930-361.

15. Ibid., 1930-362. See also H. L. Callendar, "Continuous Electrical Calorimetry," *Phil. Trans.* A, 199 (1902): 55–148.

16. "No. 15.—Professor H. L. Callendar, C.B.E.," *Middlesex County Times* (14 April 1928). "Callendar, Hugh Longbourne," *Dictionary of Scientific Biography* 3 (New York: Charles Scribner's Sons, 1971), 19–20.

17. A galvanometer is an instrument for detecting and measuring small electric currents. See London Science Museum Inventory Card, 1950-259.

18. L. H. Callendar, "H. L. Callendar—Instrument Engineer."

19. X-rays were discovered in 1895 by Wilhelm Conrad Roentgen.

20. L. H. Callendar annotations to "Obituary Notice of Hugh Longbourne Callendar," CP 8, Folder 8.

21. "No. 15.—Professor H. L. Callendar, C.B.E."

22. A. S. Eve, *Rutherford: Being the life and letters of the Rt. Hon. Lord Rutherford, O. M.* (Cambridge: Cambridge University Press, 1939); quoted in L. H. Callendar, "Professor H. L. Callendar," 2.

23. L. H. Callendar, "H. L. Callendar—Instrument Engineer."

24. Leslie Callendar to Bridget, no date, CP 8, Folder 3.

25. G. S. Callendar, 1947, draft of letter providing evidence of age, CP 8, Folder 1.

26. "The Late Professor H. L. Callendar." *Middlesex County Times* (1 February 1930): 5.

27. L. H. Callendar, "H. L. Callendar—Instrument Engineer."

28. Ibid.

29. Ibid.

30. The car was purchased by the London Science Museum in 1930 for £10 sterling. G. L. Overton to G. S. Callendar, 27 November 1930, CP 8, Folder 1. An image of the car appears in L. H. Callendar, "H. L Callendar—Instrument Engineer," 70.

31. Back of photo of 49 Grange Road, Ealing. CP 8, Folder 3.

32. *The Borough of Ealing Year Book and Who's Who* (London: Middlesex County Times Printing and Publishing Co. Ltd., 1936).

33. L. H. Callendar, "H. L. Callendar—Instrument Engineer."

34. L. H. Callendar, "Is Work Romantic?" *Circuit* (August 1956): 17–19.

35. Personal communication with Bridget, 22 December 2002.

36. L. H. Callendar, "Professor H. L. Callendar," note 7.

37. L. H. Callendar, "H. L. Callendar—Instrument Engineer."

38. Personal communication with Peter Craze, Headmaster of Durston House School, Ealing, 26 March 2004; "No. 15.—Professor H. L. Callendar, C.B.E."

39. Cyril Picciotto, *St. Paul's School* (London: Blackie and Son, 1939), 106–108; A. H. Mead, *A Miraculous Draught of Fishes: A History of St. Paul's School* (London: James and James, 1990), 101–103.

40. *St. Paul's School Registers, 1905–1985* (London, 1985); personal communication with Simon May, Archivist at St. Paul's School, London, 28 March 2002.

41. L. H. Callendar, "H. L. Callendar—Instrument Engineer," 72.

42. Mr. G. S. Callendar, Résumé, 8 November 1940, see Figure 3.4.

43. The Royal Naval Volunteer Reserves was founded in 1903 and saw extensive action in the Great War, http://www.rnr100.com (19 September 2003).

44. Hydrophones detect low frequency sounds generated by submarines. "ASDIC and Sonar Systems in the RCN," http://jproc.ca/sari/asd_gen.html (3 August 2005).

45. "Imperial College was established in 1907 as a merger of the Royal College of Science, the City and Guilds College and the Royal School of Mines." About Imperial, http://www.imperial.ac.uk/P287.htm (3 August 2005). "The City and Guilds College

forms the Engineering Section of the Imperial College of Science and Technology. It is a 'School of the University of London' in the Faculty of Engineering." *Prospectus of the City and Guilds (Engineering) College* (London: 1919–1920), 339.

46. Imperial College London, Library Archives and Special Collections, Records of the Department of Physics of Imperial College, 1882–1985, including a departmental history from 1851 to 1960.

47. H. L. Callendar to Jones, 7 November 1922, Imperial College London, Library Archives and Special Collections, KP/9/1/4. BEAMA existed to safeguard and advance the interests of British electrical manufacturers against foreign competition; http://www.BEAMA.org.uk; BEAIRA, funded by the Department of Scientific and Industrial Research, industry, and member companies, facilitated co-operative electrical research in Great Britain; http://www.era.co.uk/corporate/history.asp (3 August 2005). See also Henry G. Taylor, *An Experiment in Co-Operative Research: An Account of the First Fifty Years of the Electrical Research Association* (London: Hutchinson, 1970), 92.

48. Callendar, H. L, with G. S. Callendar, 1926 (see Appendix A, annotated bibliography). On H. L. Callendar's final work on steam see J. W. (Bill) Fox, "From Lardner to Massey: A history of physics, space science and astronomy at University College London 1826 to 1975," http://www.phys.ucl.ac.uk/department/history/BFox1.html (10 March 2006).

49. *West Middlesex Gazette*, 25 January 1930, p.10, column b. Deaths: Callendar; "The Late Professor H. L. Callendar: Famous Scientist's Career," *Middlesex County Times*, 1 February 1930, p. 5, column b.

50. Between 1942 and 1977 these royalties amounted to £5631, Longbourne to Cotching & Son, 31 January 1975, re: royalties from Callendar Steam Tables, 1 p. CP 8, Folder 1. Royalties from 1930 to 1942 are not known, but must have been substantial.

51. Ealing Lawn Tennis Association, Gentlemen's Singles Championship, http://www.ealingtennis.com/menschamps.html (17 April 2004).

52. London Science Museum, Inventory Cards, 1914-725, 1930-353, 1930-354, 1924-358, 1924-568, 1930-355, 1930-351, and 1924-567.

Chapter Two

1. See the following "Percuil" weather journals: CP 2, 1942; CP 3, 1943-ARTHUR, 1944–1945, 1948; CP 4, 1957-10; CP 6, 1960-W; CP 7, 1963-W.

2. Certified Copy of an Entry of Marriage, Guy Stewart Callendar and Phyllis Burdon Pentreath, 30 August 1930, General Register Office, England. Document ordered through Family Records Centre, London.

3. *Newspaper Press Directory* (London: Benn Booth, 1930).

4. *Middlesex County Times* (30 August 1930), p. 3, column b.

5. Personal communication with Bridget, 30 May 2005.

6. Letters from G. S. Callendar to A. C. Egerton, 20 August and 27 August 1934, Egerton Papers, B/Egerton/14/4, Imperial College London, Library Archives and Special Collections.

7. Letter from G. S. Callendar to A. C. Egerton, 26 November 1930, Egerton Papers, B/Egerton/16/2, Imperial College.

8. Their formal names were Anne Pentreath and Phyllis Bridget (who later changed her legal name to Bridget Phyllis).

9. Guy to Phyllis, September 1934, Five letters from his trip to America, CP 8, Folder 1 and Appendix B of this book.

10. "Battle of Britain Campaign Diary, 15 September 1940," http://www.raf.mod.uk/bob1940/september15.html (10 March 2006).

11. Issues of class difference during the attacks are discussed in "The Blitz: Sorting the Myth from the Reality," http://www.bbc.co.uk/history/war/wwtwo/blitz_01.shtml (10 March 2006).

12. Frank Mcc, "My Father the Firewatcher," http://www.bbc.co.uk/dna/ww2/A1250506 (10 March 2006).

13. Alan Jeffreys, "The War Through the Window of a Schoolboy," http://www.bbc.co.uk/dna/ww2/A1365743 (10 March 2006).

14. Langhurst House, Langhurstwood Road, Horsham, West Sussex RH12 4WX.

15. *Newspaper Press Directory* (London: Benn Booth, 1964).

16. The accident occurred during a test of a petrol burner.

17. Mike Brown, "Christmas Under Fire," http://www.bbc.co.uk/history/war/wwtwo/christmas_underfire_01.shtml (10 March 2006).

18. CP 2, Notebook 1942, 52–53.

19. "Horsham: General history of the town," *A History of the County of Sussex:* Volume VI Part 2: *Bramber Rape (North-Western Part) including Horsham* (1986), 131–156; http://www.british-history.ac.uk/report.asp?compid=18350 (7 July 2005).

20. Scottish novelist and dramatist Sir James Matthew (J. M.) Barrie (1860–1937) author of *Peter Pan*, and Nobel Laureate George Bernard Shaw (1856–1950).

21. Bridget currently owns a Dachshund named "Toffee."

22. See photographs of conifers dated 1956, 1953, and 1961 in CP 8, Folder 3.

23. G. S. Callendar to Anne, 11 February 1962, CP 8, Folder 1.

24. "Yes perhaps we may meet at a better time of year. I occasionally cycle Guildford way in the summer, but the weather has to be very good for it!" (Callendar to Lamb, 10 February 1962, CP 1.)

25. Before 1971, the British monetary system involved pounds (£), shillings (s), and pence (d), with 20s per pound and 12d per shilling.

26. G. S. Callendar to Anne, 1963, CP 8, Folder 1.

27. CP 6, Notebook 1960-W, 76–77.

28. CP 7, Notebook 1963-W, 44–47.

29. Certified Copy of an Entry of Death, Guy Stewart Callendar, 3 October 1964, General Register Office, England. Document ordered through Family Records Centre, London.

Chapter Three

1. BEAIRA Minute Book 2, 19 April and 23 May 1929, IMechE Library, London. "Steam-Turbine and Steam-Table Conferences: Reports of Two International Meetings in London Attended by American Engineers," *Mechanical Engineering* 51, 10 (October 1929): 790–792.

2. Nathan S. Osborne and E. F. Mueller to the Members of the London Steam Table Conference, 5 May 1930 with quote from the following enclosure: "Memorandum on the Definition of Heat Unit," 28 February 1930, Egerton Papers, B/Egerton/13/3, Imperial College. Also enclosed are documents communicating the details of this decision to other technical laboratories internationally.

3. "Steam-Turbine and Steam-Table Conferences," 792.

4. Obituary "Guy Stewart Callendar," *Quart. J. Roy. Meteor. Soc.* 91 (1965): 112.

5. BEAIRA Minute Book 2, 6 February 1930, IMechE Library.

6. *Sir Alfred Egerton, F.R.S., 1886–1959: A Memoir with Papers*, Lady Ruth Julia Egerton, Ed. (London: privately published, 1963).

7. Ibid., Preface.

8. Sir Henry Lewis Guy, F.R.S. (1887–1956), chief engineer of the mechanical department of the Metropolitan Vickers Electrical Company, 1918–1941 and Secretary for the Institution of Mechanical Engineers, 1941–1951, was very active in BEAIRA. *Dictionary of National Biography* 24 (Oxford: Oxford University Press, 2004). Sir Henry Thomas Tizard, F.R.S. (1885–1959), physical chemist and science administrator. In 1929 Tizard became rector at Imperial College, a position he maintained until 1942. *Dictionary of National Biography* 54 (Oxford: Oxford University Press, 2004). Egerton's Personal Diary, 18 February, 20 February, and 14 March 1930, GB 0117 AE/2, Egerton Papers, Royal Society of London.

9. Egerton's Personal Diary, 14 March 1930 and 26 March 1930, Royal Society of London.

10. Egerton, A. and G. S. Callendar, 1933, 149. See also Egerton's Personal Diary, 20 February 1930, Royal Society of London.

11. Egerton, A. and G. S. Callendar, 1933; quotes from 205 and 148. Frederick George Keyes (1885–1976), U.S. physical chemist, MIT professor, and coauthor, with J. H. Keenan, of *Thermodynamic Properties of Steam* (1936).

12. E. B. Wedmore to Egerton, 10 March 1930, Egerton Papers, B/Egerton/13,14, Imperial College.

13. Further details on the technical disagreements are in A. C. Egerton, "Memorandum on the Properties of Steam," 10 July 1930, Egerton Papers, B/Egerton/14/1, Imperial College.

14. Sources used in this chapter include Guy to Phyllis, September 1934, five letters from trip to America, 25 p. CP 8, Folder 1 (and Appendix B of this book); A. Egerton, "[Journal of a trip to] America, September 1934," AE/3/5, Egerton Papers, Royal Society of London; Mrs. A. Egerton, "Trip to America, September 1934," AE/4/1, Egerton Papers, Royal Society of London; "Third International Steam Table Conference, General Program. September 17 to 22, 1934," *The American Society of Mechanical Engineers*, copy in Egerton Papers, B/Egerton/14/2, Imperial College; and "Third International Steam Table Conference, September 17–22." *Mechanical Engineering* 56, 11 (1934): 701–703.

15. Henry Ford (1863–1947), automobile manufacturer and entrepreneur, *American National Biography* 8 (New York, Oxford: Oxford University Press, 1999), 226–235.

16. The *Britannic*, a 27,000-ton vessel with a 683 ft 8 in keel, 82 ft 6 in beam, and a cruising speed of 18 knots, was the third ship with this name. All three were built by Harland and Wolff for the White Star Line. White Star Fleet List, http://www.simplonpc.co.uk/WhiteStar3.html (4 August 2005); "The Cunard White Star Motorship Britannic," http://www.uncommonjourneys.com/pages/britannic (4 August 2005).

17. Guy to Phyllis, 8–9 September 1934, CP 8, Folder 1 (and Appendix B).

18. "White Star Line," http://www.red duster.co.uk/WSTAR11.htm (11 March 2006).

19. Guy to Phyllis, 8–9 September 1934, CP 8, Folder 1 (and Appendix B).

20. Guy to Phyllis, 20–21 September 1934, CP 8, Folder 1 (and Appendix B).

21. CP 8, Folder 3.

22. Mrs. A. Egerton, "Trip to America."

23. Guy to Phyllis, 20–21 September 1934, CP 8, Folder 1 (and Appendix B).

24. "Third International Steam Table Conference," 1934.

25. Davis played an integral organizational role in organizing the first ASME meeting on the thermodynamic properties of water and steam in 1921. "Brief History of the ASME Properties of Steam Subcommittee, http://www.asme.org/research/wsts/steam_history.pdf (4 August 2005).

26. "Third International Steam Table Conference, General Program." The *New York*, a 21 455-ton trans-Atlantic liner, was built in 1926.

27. Mrs. A. Egerton, "Trip to America," 8–9. "Shipping and Mails," *New York Times* (20 September 1934), p. 47.

28. The Astor Hotel, noted for its enormous public rooms and an elaborate roof garden, was built in 1904 at Broadway and 44th Street. Ruth met Professors Ernst Schmidt of Danzig, Fritz Henning of Berlin, Werner Koch of Munich, and Erich J. M. Honigmann of Vienna.

29. Guy to Phyllis, 20–21 September 1934, CP 8, Folder 1 (and Appendix B).

30. "Third International Steam Table Conference," 1934.

31. "Third International Steam Table Conference, General Program."

32. Guy to Phyllis, 22 September 1934, CP 8, Folder 1 (and Appendix B). "Third International Steam Table Conference," 1934.

33. Guy to Phyllis, 22 September 1934, CP 8, Folder 1 (and Appendix B).

34. Keenan, who had served as a delegate to earlier international steam table conferences, was the co-author, with F. G. Keyes, of the *Thermodynamic Properties of Steam* (1936).

35. "Steamers Crash in Fog Off Cape; Liner Laconia Rips Hole in Side of Freighter Pan Royal Off Peaked Hill Bar," *New York Times* (25 September 1934), p. 45. "Laconia Out for Repairs," *New York Times* (27 September 1934), p. 45.

36. "Shipping and Mails," *New York Times* (24 September 1934), p. 35. The *Mauretania*, built in 1907 as the largest ship in the world, was a 31 938-ton liner with a length of 760 ft and a beam of 87.5 ft. An older sister of the *Lusitania*, she served as a troop transport and hospital ship in World War I and, after being damaged by fire, was converted from coal to oil in 1921. On 26 September 1934 the *Mauretania* left New York on her final Atlantic crossing. Cunard Line, Page 2—Ocean Liners 1900–1914, http://www.simplonpc.co.uk/Cunard2.html#anchor424250 (11 March 2006).

37. Callendar, G. S., Hugh L. Callendar, and Sir A. C. Egerton, 1939; and Egerton, A. C. and G. S. Callendar, 1939.

38. A. C. Egerton to D. V. Onslow, 11 January 1939, Egerton Papers, B/Egerton/13/2, Imperial College.

39. "Memorandum on E.R.A. Steam Research, September 1939," Egerton Papers, B/Egerton/13/2, Imperial College.

40. BEAIRA Minute Book 3, Section J: Steam Power Plant, Meeting on 17 May 1940, J/M26, 1-6, pp. 156–161, IMechE Library.

41. BEAIRA Minute Book 3, Section J: Steam Power Plant, Meeting on 11 July 1941, J/M27, 1-5, pp. 178–182, IMechE Library.

42. H. L. Guy to E. B. Wedmore, 5 August 1940 (confidential copy to Egerton), Egerton Papers, B/Egerton/13/2, Imperial College.

43. E. B. Wedmore to H. L. Guy, August 1940 (confidential copy for Egerton enclosed with H. L. Guy to A. C. Egerton, 27 August 1940), Egerton Papers, B/Egerton/13/2, Imperial College.

44. H. L. Guy to Egerton, 27 August 1940.

45. H. L. Guy to E. B. Wedmore, 14 October 1940, copy in the Egerton Papers, B/Egerton/13/2, Imperial College.

46. *Sir Alfred Egerton, F.R.S.*, 1886–1959, 44–45.

47. Callendar's résumé, 8 November 1940, Egerton Papers, B/Egerton/13/2, Imperial College.

48. Wedmore to Egerton, 31 October 1941 and Egerton to Wedmore 4 November 1941, Egerton Papers, B/Egerton/13/2, Imperial College.

49. Callendar, G. S., and Sir Alfred Egerton, 1960.

50. Egerton to D. C. Martin, 28 January 1959, Egerton Papers, B/Egerton/19, Imperial College.

51. Egerton's Diary, October 1958, AE/2/37, vol. x, series c, p. 82, Royal Society of London.

Chapter Four

1. Obituary, "Guy Stewart Callendar."

2. R. G. H. Watson, Electrochemical Generation of Electricity," *Research* 7, 1 (January 1954): 34–40; reference from CP 7, Notebook References, 88. This article traces the history of fuel cells from Volta's principle of the electrochemical generation of electricity to contemporary work.

3. N. Sheppard, "Gordon Brims Black McIvor Sutherland," *Biogr. Mem. Fellows Roy. Soc.* 28 (November 1982): 588–562; quote from 601.

4. Callendar, G. S., 1941c.

5. Sutherland, G. B. B. M., and G. S. Callendar, 1942–43. See CP 1 for exchange of letters with Sutherland in 1948 and CP 2, Notebook 1942-IRS.

6. Egerton was well aware of this work, as evidenced by extensive notes in his diary. Egerton Personal Diary, 1943–44, AE/2/4, pp. 69–70, Royal Society of London.

7. The Ministry of Supply was formed in 1939 to coordinate the provision of equipment to the British armed forces. It was amalgamated with the Ministry of Aircraft Production in 1946 and was superseded by other agencies in 1959. The Armament Research Department (ARD) was established in 1942 and was amalgamated with the design department in 1955 to form the Armament Research and Development Establishment (ARDE).

8. Petroleum Warfare Department Folder, SUPP 15/2; and memo from F. A. Gear, 15 May 1942, AVIA 22/2303, British National Archives.

9. "A. C. Hartley," *Dictionary of National Biography* (Oxford: Oxford University Press, 2004).

10. Joint press release from Petroleum Warfare Department and Air Ministry, FIDO Conference Program, 30 May 1945, CP 8, Folder 6 (hereafter referred to as FIDO Conference Program).

11. Ibid.

12. Records of the Petroleum Warfare Department (see note 8 above).

13. Sir Donald Banks, *Flame Over Britain: A Personal Narrative of Petroleum Warfare* (London: Sampson Low, Marston & Co., 1946), 2, 139.

14. Geoffrey Williams, *Flying Through Fire: The Fogbuster of World War Two* (London: Grange, 1996), 1–6.

15. Patent Application, 24 February 1945, for "Fog freeable runways for aircraft and plant associated therewith," CP 8, Folder 6.

16. Williams, *Flying Through Fire*, 6.

17. Ibid., 7.

18. G. I. Taylor, "Fog Conditions," *Aero. J.*, 21 (1917): 75.

19. Napier Shaw, "Artificial Dissipation of Fog," *Meteor. Mag.* 55 (January 1921): 265–268. See also Royal Aircraft Establishment and Meteorological Office, "Fog and Its Dispersion," *Reports and Memoranda, Aeronautical Research Committee*, No. 727 (April, 1921).

20. F. A. Lindemann, "Notes on Landing in Fog," *Reports and Memoranda, Aeronautical Research Committee*, No. 726 (February 1921), cited in E. G. Walker and D. A. Fox, *The Dispersal of Fog from Airfield Runways* (London: Ministry of Supply, 1946), 3.

21. C. Morgan, "Fog continues to battle scientists," *Tycos* 19 (January 1929): 33; and Alexander McAdie, "The Control of Fog," *Sci. Mon.* 33, 1 (July, 1931): 28–36.

22. W. J. Humphreys, *Rain Making and Other Weather Vagaries* (Baltimore: Williams & Wilkins, 1926).

23. D. Brunt, "The Artificial Dissipation of Fog," *J. Sci. Instruments* 16, 5 (May 1939): 137–140.

24. R. J. Ogden, "Fog Dispersal at Airfields," *Weather* 43 (1998): 20–23, 34–38.

25. Callendar's résumé, 1940, see Figure 3.4.

26. FIDO Conference Program.

27. Banks, *Flame Over Britain*, 139.

28. Ibid.

29. Ibid.

30. FIDO Conference Program.

31. Ibid.

32. Ogden, "Fog Dispersal at Airfields," 21.

33. Ibid.

34. Banks, *Flame Over Britain*, 148–50; emphasis added.

35. For an aerial view of FIDO in use at Fiskerton on 3 November 1943, see Royal Air Force Museum photo PC71/19/524.

36. Williams, *Flying Through Fire*, 20.

37. Banks, *Flame Over Britain*.

38. CP 2, Notebook 1942-IRS, 217 and CP 8, Folder 6, G. S. Callendar, "List of own reports for P.W.R.S. 1942–46," dated March 1946.

39. Notes on FIDO Meetings, 12 July 1943, Lord Geoffrey Lloyd Box, Folder 85/46/18, Imperial War Museum. A. C. Hartley, "Fog Dispersal," *J. Roy. Soc. Arts* (6 December 1946): 35–37.

40. A. O. Rankine, "FIDO Investigation Wind Tunnel Experiments," Petroleum Warfare Department, 4 April 1945, p. 42, AIR 20/7239, British National Archives; and A. O. Rankine, "Experiments in the Empress Hall, Earl's Court, London, on the distribution of heat from line burners in relation to the problem of fog clearance in airfields," Petroleum Warfare Department, 1943 (copy in Imperial War Museum). FIDO Conference Program; Williams, *Flying Through Fire.*

41. Walker and Fox, *Dispersal of Fog*, 1–9, 33–36.

42. "Mr. A. C. Hartley's Diary, 30 September 1942 – 2 August 1943," and "FIDO Statement by A. C. Hartley, February 17, 1950," Lord Geoffrey-Lloyd Box, Imperial War Museum.

43. Banks, *Flame Over Britain*, 153.

44. CP 8, Folder 6, "List of own reports for P.W.R.S., 1942–46."

45. Mr. A. C. Hartley's Diary, 30 September 1942 – 2 August 1943.

46. "Notes on FIDO Meetings," 12 July 1943 and 14 March 1944, Lord Geoffrey Lloyd Box, Folder 85/46/18, Imperial War Museum.

47. FIDO Conference Program.

48. Ogden, "Fog Dispersal at Airfields," 34.

49. Ibid.

50. "Remarks by pilots . . ., Do you know FIDO?" Sir Donald Banks Box, Imperial War Museum.

51. Williams, *Flying Through Fire*, 39.

52. Ogden, "Fog Dispersal at Airfields," 35–37. On American-built FIDO systems see Dave Zebo, Thermal Fog-Dispersal, High Pressure System (1957), http://www.northcoast.com/~bbn/zebo.html (30 January 2006)

53. FIDO Conference Program; and "FIDO Principle Press Cuttings," Sir Donald Banks Box, Imperial War Museum.

54. For contemporary accounts of FIDO by airmen see "Fido No Fog," *Tee Emm* 4 (May 1944): 35 and "Ninth Time Lucky," *Tee Emm* 4 (October 1944): 155–56. Copy of the latter in *Churchill at War: The Prime Minister's Office Papers, 1940–1945*, PRO Record Classes PREM 3 and PREM 4, microfilm edition (Berkshire: Primary Source Media, 1998), reel 2.

55. Ibid.

56. Ibid.

57. Total fuel use in all of World War II has been estimated at 7 billion barrels of oil, or the equivalent of 150 billion gallons of gasoline; Keith Miller, "How Important

Was Oil in World War II?" http://hnn.us/articles/339.html (11 March 2006). M. Garbett and B. Goulding, *The Lancaster at War* (London: Ian Allen, 1971). Other estimates appear in Tom Morrison, *Quest for All-Weather Flight* (Shewsbury, UK: Airlife, 2002), 155–157.

58. Jean-Pierre Chalon, "Rapporteur on fog dissipation," World Meteorological Organization, CAS Working Group on Physics and Chemistry of Clouds and Weather Modification Research, 21st Session, Geneva, 23–27 May 2005. PCCWMR-21/Doc. 4.3. (17.V.2005).

59. Ogden, "Fog Dispersal at Airfields," 38.

60. CP 8, Folder 6, "List of own reports for P.W.R.S., 1942–46."

61. CP 8, Folder 6, "Experiments on the Thermal Cutting of Wood," 30 June 1944.

62. CP 8, Folder 6, G. S. Callendar, "Report on tests with anti-surge baffles fitted in a tank," ca. 1940s.

63. CP 8, Folder 6, "Local Team Helps to Save Beaches," *West Sussex County Times* (7 April 1967).

64. CP 8, Folder 7, G. S. Callendar, "Test of Daniell's 'Dragon' Heater in West Hangar, 10th October, 1950," and "Trial of a self contained portable space heating unit," Report A.D.E. 3/51, February 1951.

65. CP 8, Folder 7, G. S. Callendar, "The diffusion of high pressure air into liquids through flexible membranes," April 1953.

66. G. S. Callendar, 1956, "Gravity Method of Obtaining a Low Pressure High Velocity Air Current for Laboratory Research," Ministry of Supply Branch Memorandum S4/8/56.

67. A venturi tube, named for the Italian physicist G. B. Venturi (1746–1822), is a short pipe with a constricted inner surface. In Callendar's design, fluid passing through the tube speeds up as it enters the orifice and the pressure drops, acting as further source of suction.

68. Callendar's salary was £550 per annum in 1942 according to "List of Staff at Research Department, Langhurst," AVIA 22/2303, British National Archives.

69. CP 8, Folder 1, V. M. Callendar, 25 November 1947, sworn affidavit re: birth of G. S. Callendar.

Chapter Five

1. CP 1, Callendar to Gilbert Plass, 13 May 1957.

2. A complete list of Callendar's climate publications appears in the annotated bibliography.

3. CP 2, Notebook 1942-IRS, 8.

4. For extended treatments of these themes see Fleming, *Historical Perspectives on Climate Change.*

5. I. Grattan-Guinness, with J. Ravitz, *Joseph Fourier, 1768–1830: A Survey of his Life and Work, Based on a Critical Edition of his Monograph on the Propagation of Heat Presented to the Institute of France in 1807* (Cambridge: MIT Press, 1972); and John Herivel, *Joseph Fourier: The Man and the Physicist* (Oxford: Clarendon Press, 1975).

6. Joseph Fourier, cited in Fleming, *Historical Perspectives on Climate Change,* p. 153, note 36.

7. Joseph Fourier, "Remarques générales sur les températures du globe terrestre et des espaces planétaires," *Ann. Chim. Phys.*, 2nd series, 27 (1824): 136–67. English translation by Ebeneser Burgess in *Amer. J. Sci. Arts* 32 (1837): 1–20.

8. Ibid., 155; Burgess translation, 13.

9. Ibid., 151–53; Burgess translation, 10–11.

10. John Tyndall, "On the Transmission of Heat of Different Qualities Through Gases of Different Kinds," *Proc. Roy. Inst. Gt. Br.* 3 (1858–1862): 158.

11. John Tyndall, "On Radiation through the Earth's Atmosphere," Friday, 23 January 1863, *Proc. Roy. Inst. Gt. Br.* 4 (1851–1866): 4–8; quote from 8; also in *Phil. Mag.*, Series 4, 25 (1863): 200–206.

12. W. F. Barrett, "On a Physical Analysis of the Human Breath," *Phil. Mag.* 28 (1864): 108–121. This article directly follows articles on radiant heat by Tyndall and on glacial climates by James Croll.

13. Fleming, *Historical Perspectives on Climate Change,* 68–71.

14. Svante Arrhenius, "On the Influence of Carbonic Acid in the Air upon the Temperature of the Ground," *Phil. Mag.*, Series 5 (1896): 237–276; Elisabeth Crawford, *Arrhenius: From Ionic Theory to the Greenhouse Effect* (Canton, MA: Science History Publications, 1996); Fleming, *Historical Perspectives on Climate Change,* 74–82.

15. James Rodger Fleming, "Global Climate Change and Human Agency: Inadvertent Influence and 'Archimedean' Interventions," *Intimate Universality: Local and Global Themes in the History of Weather and Climate,* James Rodger Fleming, Vladimir Jankovic, and Deborah R. Coen, Eds. (Sagamore Beach, Mass.: Science History Publications/USA, 2006), 223–248.

16. Nils Ekholm, "On the Variations of the Climate of the Geological and Historical Past and Their Causes," *Quart. J. Roy. Meteor. Soc.* 27 (1901): 61. This article appeared in Swedish in 1899 and was translated two years later.

17. Svante Arrhenius, *Worlds in the Making: The Evolution of the Universe,* translated by H. Borns (New York: Harper, 1908), 54–63; Fleming, *Historical Perspectives on Climate Change,* 82, 111; David Keith, "Geoengineering & Climate: An Overview," unpublished paper presented at the Tyndall Centre conference on Macro-engineering

Options for Climate Change Management & Mitigation, Cambridge, UK, January 2004.

18. Richard Joel Russell, "Climatic Change through the Ages," in U.S. Department of Agriculture, *Climate and Man: Yearbook of Agriculture 1941*, House Document 27, 77th Congress, 1st session (Washington, D.C., 1941), 67–97, quote from 94.

19. T. C. Chamberlin, "A Group of Hypotheses Bearing on Climatic Changes," *J. Geol.* 5 (1897): 653–683; and Chamberlin, "An Attempt to Frame a Working Hypothesis of the Cause of Glacial Periods on an Atmospheric Basis," *J. Geol.* 7 (1899): 545–584, 667–685, 751–787.

20. Knut Ångström, "Ueber die Bedeutung des Wasserdampfes und der Kohlensäure bei der Absorption der Erdatmosphäre," *Ann. Phys. Chimie* 5, 324, (1900): 720–733; and Ångström, *Ann. Phys. Chimie* 6 (1901): 690. CP 2, Notebook 1942-IRS, 12.

21. C. G. Abbot and F. E. Fowle, *Annals of the Astrophysical Observatory*, Smithsonian Institution, Vol. 2 (1908): 172.

22. G. C. Simpson, "Past Climates," *Manchester Lit. Philos. Soc. Mem.* 74, 1 (1929–30): 9–10.

23. Many of these theories are surveyed in C. E. P. Brooks, *Climate Through the Ages: A Study of the Climatic Factors and Their Variations*, 2nd ed., rev. (New York: McGraw-Hill, 1949). See also Brooks, "Selective Annotated Bibliography on Climatic Changes," *Meteorological Abstracts and Bibliography* 1, 4 (1950): 446–475.

24. W. J. Humphreys, "Volcanic Dust and Other Factors in the Production of Climatic Changes, and Their Possible Relation to Ice Ages," *J. Frankl. I.* 176 (1913): 132.

25. CP 2, Notebook 1942-IRS, 14 and 195; Callendar was referring to N. Ekholm, *Meteor. Z.* 19 (1902):491.

26. L. Paschen, 1892, *Ann. Phys. Chemie* 51; CP 2, Journal 1942-IRS, 10.

27. E. O. Hulburt, 1931, "The Temperature of the Lower Atmosphere of the Earth." *Phys. Rev.* 38 (1876–90); CP 2, Notebook 1942-IRS, 197.

28. Callendar, G. S., 1938a.

29. CP 2, Notebook 1942, 257.

30. CP 8, Folder 4.

31. Callendar, G. S., 1938a.

32. Callendar, G. S., 1938b,c,d and CP 2, Notebook 1942, 256-57.

33. Callendar, G. S., 1939a. Callendar, 1939c is a short entertaining piece.

34. Callendar's résumé, 1940, see Figure 3.4.

35. Callendar, G. S., 1939b.

36. Ibid. Mudge, F. B., "The Development of the 'Greenhouse' Theory of Global Climate Change from Victorian Times," *Weather* 52 (1997): 13–17. The author claims that inaccuracies in the measurement of CO_2 were "greatly reduced" by 1900 and tended to cluster around 300 parts per million. From and Keeling disagree, see note 15, chapter 6.

37. Callendar, G. S., 1940.

38. Callendar, G. S., 1941c.

39. Wexler, H., *Mon. Wea. Rev.* 64 (1936): 122; Elsasser, W. M., *Mon. Wea. Rev.* 66 (1938): 175; Adel, A. and V. Slipher, *Astrophys. J.* 89 (1939): 21.

40. Callendar, G. S., 1941c.

41. Callendar, G. S., 1941d, 1941e, and 1942.

42. Sutherland, G. B. B. M. and G. S. Callendar, 1942–43.

43. Callendar, G. S., 1943, 1944.

44. Callendar, G. S., 1947–48.

45. Callendar, G. S., 1949.

46. Callendar, G. S., 1950a,b, 1951a,b, 1952a,b, and 1955.

47. Callendar, G. S., 1957a; CP 1, Plass to Callendar, 5 April 1957; CP2, Notebook 1942, 269.

48. CP 2, Notebook 1942, 265.

49. H. W. Ahlmann, "The Present Climatic Fluctuation," *Geogr. J.* 112 (October–December 1948): 165–95.

50. H. C. Willett, "Temperature Trends of the Past Century," *Centennial Proc. Roy. Meteor. Soc.* (1950), 195–211.

51. J. O. Fletcher, "Climatic Change and Ice Extent on the Sea," RAND Paper P-3831 (April 1968).

52. "Getting Warmer?" *Time Magazine* (15 May 1950).

53. Albert Abarbanel and Thorp McClusky, "Is the World Getting Warmer?" *Saturday Evening Post* (1 July 1950): 22–23, 57, 60–63.

54. "Invisible Blanket," *Time Magazine* (25 May 1953).

55. Engel, Leonard, "The Weather Is *Really* Changing," *New York Times Magazine* (12 July 1953): 7ff. ProQuest Historical Newspapers The New York Times (1851–2002); See also William J. Baxter, *Today's Revolution in Weather!*, cited in Fleming, 1998.

56. "One Big Greenhouse," *Time Magazine* (28 May 1956): 59. The article cited annual world CO_2 emissions of 0.5 billion tons in 1860, 9 billion tons in 1950, and a projected 47 billion tons in 2010. The current projection for 2010 is just under 28 billion tons.

57. A basic description of the IGY can be found at U.S. National Academies, "The International Geophysical Year," http://www.nas.edu/history/igy (accessed 28 July 2006). The author (Fleming) is currently involved in historical reassessments of the social and intellectual implications of all the International Polar Years.

58. CP 1, Callendar to Plass, 12 February 1958; Keeling to Callendar, 5 February 1958; and Callendar to Keeling, 12 February 1958.

59. Suess, Hans E., 1953, "Natural Radiocarbon and the Rate of Exchange of Carbon Dioxide Between the Atmosphere and the Sea," *Proc. Conf. Nuclear Processes in Geologic Settings* (Williams Bay, Wisconsin, 21–23 September 1953), 52.

60. Appearing in the same issue of *Tellus* 9 (1957) are Harmon Craig, "The Natural Distribution of Radiocarbon and the Exchange Time of Carbon Dioxide Between Atmosphere and Sea," 1–17; Roger Revelle and Hans E. Suess, "Carbon Dioxide Exchange between Atmosphere and Ocean and the Question of an Increase in Atmospheric CO_2 during the Past Decades," 18–27; and James R. Arnold and Ernest C. Anderson, "The Distribution of Carbon-14 in Nature," 28–32.

61. Plass, G. N., "The Carbon Dioxide Theory of Climatic Change," *Tellus* 8 (1956): 140–54 and Dingle, H. N., "The Carbon Dioxide Exchange between the North Atlantic Ocean and the Atmosphere," *Tellus*, 6 (1954): 342.

62. Mark Bowen, *Thin Ice* (New York: Henry Holt, 2005), 109–110. Bowen's chapters 6 and 7 follow the outline of Fleming, *Historical Perspectives on Climate Change*. For valuable perspectives on Revelle, see Ronald Rainger, "'A Wonderful Oceanographic Tool': The Atomic Bomb, Radioactivity and the Development of American Oceanography," *The Machine in Neptune's Garden: Historical Perspectives on Technology and the Marine Environment*, Helen M. Rozwadowski and David K. Van Keuren, Eds. (Sagamore Beach, Mass.: Science History Publications/USA, 2004), 96–132.

63. CP 3, Notebook 1956-11, 28-32; CP 1, Callendar to Plass, 13 May 1957; Callendar, G. S., 1957a; Callendar, G. S., 1958a; and CP 2, Notebook 1942, 269.

64. CP 1, Plass to Callendar, 30 December 1957; Callendar 1957b.

65. In Bert Bolin, Ed., *The Atmosphere and the Sea in Motion: Scientific Contributions to the Rossby Memorial Volume* (New York: Rockefeller Institute Press, 1958), 130–142.

66. The 96 notebooks, including 25 with overlapping content, have the following themes: British temperatures 7%, CO_2 11%, glaciers 5%, infrared 8%, Percuil weather data 6%, references 6%, sea temperatures 1%, and world and regional temperatures 68%.

67. Callendar, G. S., 1961a,b. CP 1 contains related correspondence with Landsberg and Mitchell. Draft copy of the article and data in CP 8, Folder 2, Notebook 1960-05-17; reprint and retrospective note in CP 8, Folder 4; comments in CP 2, Notebook 1942, 271.

68. CP 1, Levinson, 19 November 1960.

69. CP 8, Folder 4, 1964. Temperature note 1, 2.

70. Callendar, G. S., 1964.

71. Hubert H. Lamb, "Some Aspects of the Cold, Disturbed Climate of Recent Centuries, the 'Little Ice Age,' and Similar Occurrences," *Pure and Applied Geophysics* 119 (1980/81): 628–629.

72. CP 1, Gordon Manley to Callendar, 8 October 1958.

73. Bowen, *Thin Ice*, 96.

Chapter Six

1. *West Sussex County Times* (9 October 1964), p. 24, column 1. See also Certified Copy of an Entry of Death, Guy Stewart Callendar, 3 October 1964, General Register Office, England. Document ordered through Family Records Centre, London.

2. CP 8, Folder 1, R. C. Thrush to Phyllis, 21 October 1964.

3. CP 8, Folder 1, Gilbert Plass to Phyllis, 9 November 1964.

4. CP 8, Folder 1, Gordon Manley to Phyllis, 18 October 1964.

5. CP 8, Folder 1, Derik Schove to Phyllis, 12 October 1964.

6. CP 8, Folder 5, Red Notebook.

7. CP 1, Callendar_n.d._charts_oceanCO$_2$

8. CP 1, Calledar_n.d._ms_CO$_2$spectrum.

9. CP 8, Folder 1, W. Thompson to Anne, 12 December 1964. Thompson, a former colleague of Callendar, offered £15 15s for a calculator, set of mechanical drawing instruments, micrometer caliper, vernier gauge, try and miter square with spirit level, and a small brass microscope. The complete list of instruments is in CP 8, Folder 5, Red Notebook, 13.

10. CP 8, Folder 1, Royal Society to Anne, 6 January 1965 and 11 January 1965.

11. CP 8, Folder 1, Anne to Schove, 11 December 1964.

12. "Bibliography" [of the writings of D. J. Schove], 2 p. typescript, 1985, copy in Wellcome Institute Library, London.

13. Machta (1919–2001) supported ongoing efforts, begun during the IGY, to measure background carbon dioxide concentrations and developed NOAA's Geophysical Monitoring for Climate Change network which monitors the composition of the Earth's atmosphere; *EOS* (30 October 2001): 515.

14. CP 8, Folder 1, Lester Machta to P. Goldsmith, 30 March 1979; P. Goldsmith to Anne, 25 April and 8 May 1979.

15. CP 2, Notebook 1939-40-T1, "Observations on the amount of carbon dioxide in the air. Manuscript Notebook, dated "1939-40", 136 p. CP 2 (with notebook). Charles D. Keeling to James Fleming, 9 February 2005. Keeling writes, "I also recall with pleasure a visit to Justin Schove who found the notebook for me in a cardboard box of items in London." See also Eric From and Charles D. Keeling, "Reassessment of late 19th century atmospheric carbon dioxide variations in the air of western Europe and the British Isles based on an unpublished analysis of contemporary air masses by G. S. Callendar," *Tellus* 38B (1986): 87-105.

16. CP 8, Folder 1, Schove to Anne, 29 January 1984.

17. CP 8, Folder 1, Anne to Schove, 5 February 1984.

18. CP 8, Folder 1, Schove to Archivist, 24 April 1984; C. Rawlins to Anne, 4 November 1986; with reply 8 November 1986; Mary Dobson to Anne, 7 January 1987; Eric Harris to Anne, 2 February 1987; Harris to Anne, 27 February 1989.

19. CP 8, Folder 1, Tom Wigley to Anne, 3 March 1989.

20. CP 8, Folder 1, Phil Jones to James Fleming, 26 March 2002

21. *The Papers of Guy Stewart Callendar*, Digital Edition on DVD, edited, compiled, arranged, indexed, and scanned by James Rodger Fleming and Jason Thomas Fleming (Boston: American Meteorological Society, 2007). Considerable effort has been made to ensure that the organization of the eight boxes of Callendar Papers in the CRU Library match the digital files. Fleming was granted permission to quote, cite, reproduce, and otherwise use and disseminate the information and images from this collection by both the CRU (CP 8, Folder 1, Phil Jones to James Fleming, 14 March 2003) and Callendar's daughter (CP 8, Folder 1, Bridget to James Fleming, 1 May 2003).

22. A. G. Gaydon, "Sir Alfred Egerton's Scientific Work," in *Sir Alfred Egerton, F.R.S., 1886–1959*, 198.

23. Taylor, *Experiment in Co-operative Research*, 92.

24. Gordon Manley, "Some Recent Contributions to the Study of Climatic Change," *Quart. J. Roy. Meteor. Soc.* 70 (1944): 197–219.

25. Harry Wexler, "Recent Progress in the Investigation of the Composition and Structure of the Atmosphere," Symposium on Progress in Meteorological Research, 30th Anniversary Meeting of the American Meteorological Society, St. Louis, MO, 4–6 January 1950, manuscript notes in Library of Congress, Harry Wexler Papers, Box 16, Folder Speeches and Lectures, 1950. Wexler returned to this theme in 1952: "The Radiation Balance of the Earth as a Factor in Climatic Change," Symposium on Climatic Change, American Academy of Arts and Sciences, Boston, May 9–10, 1952. See idem. Folder Speeches and Lectures, 1952.

26. CP 8, Folder 1, Schove to Bridget, 6 December 1974.

27. CP 8, Folder 1, Ferren MacIntyre to Callendar, 5 July 1977.

28. Wigley, T. M. L., P. D. Jones, and P. M. Kelly, "Warm World Scenarios and the Detection of Climatic Change Induced by Radiatively Active Gases," in SCOPE 29—The Greenhouse Effect, Climatic Change, and Ecosystem, no date, but ca. 1985, http://www.icsu-scope.org/downloadpubs/scope29/chapter06.html (21 April 2004); see also Jones, P. D., T. M. L. Wigley, and P. M. Kelly, "Variations in surface air temperature: Part 1, Northern Hemisphere 1881–1980," *Mon. Wea. Rev.* 110 (1982): 59–70.

29. CP 2, Notebook 1939-40-T1, Charles D. Keeling to James Fleming, 9 February 2005.

30. M. D. Handel and J. S. Risbey, "An Annotated Bibliography on the Greenhouse Effect and Climate Change," *Climatic Change* 21 (1992): 97–255.

31. Spencer Weart, "The Discovery of Global Warming," http://www.physicist.org/history/climate/co2.htm (7 February 2006). This is based on Weart, "From the

Nuclear Frying Pan into the Global Fire," *Bull. Atom. Sci.* (June 1992): 19–27, which claims that Callendar's work was obscure and no one really cared, and Weart, "Global Warming, Cold War, and the Evolution of Research Plans," *Hist. Stud. Phys. Sci.* 27 (1997): 319–56, which emphasizes Callendar's obscurity and amateur status.

32. Peter Brimblecombe and Ian Langford, "Guy Stewart Callendar and the Increase in Global Carbon Dioxide," *Air and Waste Management Association* (1995): 95-WA74A.02.

33. Fleming, *Historical Perspectives on Climate Change*; CP 8, Folder 1, Bridget to James Fleming, 17 December 2002. Being "chuffed" is being generally happy with life, "English to American Dictionary," http://english2american.com/dictionary/ c.html#chuffed (11 March 2006).

34. Bowen, *Thin Ice*, 96.

Index